Effects of Information and Communications Technology on Travel Choices 2018

Transportation Research Record: Journal of the Transportation Research Board

Volume 2672, Issue 50

ISBN: 978-1-5108-7838-9

Printed from e-media with permission by:

Curran Associates, Inc.
57 Morehouse Lane
Red Hook, NY 12571

proceedings.com

Some format issues inherent in the e-media version may also appear in this print version.

Published by Sage Publications USA

ISBN (Print) 978-1-5108-7838-9
ISBN 2018 Printed Set (All Issues) 978-1-5108-7735-1

Printed by Curran Associates, Inc. (2019)

Additional copies of this publication are available from:

Curran Associates, Inc.
57 Morehouse Lane
Red Hook, NY 12571 USA
Phone: 845-758-0400
Fax: 845-758-2633
Email: curran@proceedings.com
Web: www.proceedings.com

Volume 2672 Number 50 December 2018

TRR
JOURNAL OF THE TRANSPORTATION RESEARCH BOARD

Contents

Effects of Information and Communications Technology on Travel Choices

Articles

TRANSPORTATION RESEARCH BOARD

CORE SPONSORS

Transportation Departments of the 50 States and the District of Columbia

Federal Government

U.S. Department of Transportation
- Federal Aviation Administration (FAA)
- Federal Highway Administration
- Federal Motor Carrier Safety Administration (FMCSA)
- Federal Railroad Administration (FRA)
- Federal Transit Administration
- National Highway Traffic Safety Administration (NHTSA)
- Office of the Assistant Secretary of Transportation for Research and Technology (OST-R)

U.S. Air Force Civil Engineer Center (AFCEC)
U.S. Coast Guard (USCG)
U.S. Department of Energy
U.S. Department of the Interior (DOI)
U.S. Environmental Protection Agency (EPA)

Nongovernmental Organizations

American Public Transportation Association (APTA)
Association of American Railroads (AAR)
California Air Resources Board

The National Academies of
SCIENCES · ENGINEERING · MEDICINE

The **National Academy of Sciences** was established in 1863 by an Act of Congress, signed by President Lincoln, as a private, nongovernmental institution to advise the nation on issues related to science and technology. Members are elected by their peers for outstanding contributions to research. Dr. Marcia McNutt is president.

The **National Academy of Engineering** was established in 1964 under the charter of the National Academy of Sciences to bring the practices of engineering to advising the nation. Members are elected by their peers for extraordinary contributions to engineering. Dr. C. D. Mote, Jr., is president.

The **National Academy of Medicine** (formerly the Institute of Medicine) was established in 1970 under the charter of the National Academy of Sciences to advise the nation on medical and health issues. Members are elected by their peers for distinguished contributions to medicine and health. Dr. Victor J. Dzau is president.

The three Academies work together as the National Academies of Sciences, Engineering, and Medicine to provide independent, objective analysis and advice to the nation and conduct other activities to solve complex problems and inform public policy decisions. The Academies also encourage education and research, recognize outstanding contributions to knowledge, and increase public understanding in matters of science, engineering, and medicine.

Learn more about the National Academies of Sciences, Engineering, and Medicine at **www.national-academies.org.**

The **Transportation Research Board** is one of seven major programs of the National Academies of Sciences, Engineering, and Medicine. The mission of the Transportation Research Board is to increase the benefits that transportation contributes to society by providing leadership in transportation innovation and progress through research and information exchange, conducted within a setting that is objective, interdisciplinary, and multimodal. The Board's varied activities annually engage about 7,000 engineers, scientists, and other transportation researchers and practitioners from the public and private sectors and academia, all of whom contribute their expertise in the public interest. The program is supported by state transportation departments, federal agencies including the component administrations of the U.S. Department of Transportation, and other organizations and individuals interested in the development of transportation.

Learn more about the Transportation Research Board at **www.TRB.org**.

TRR
JOURNAL OF THE TRANSPORTATION RESEARCH BOARD

Article

Transportation Research Record
2018, Vol. 2672(50) 1–10

DOI: 10.1177/0361198118794044
journals.sagepub.com/home/trr
SAGE

Accessibility or Innovation? Store Shopping Trips versus Online Shopping

Kees Maat[1] and Rob Konings[1]

Abstract

The increasing penetration of online shopping will have major effects on physical stores. And the question is: In which areas will consumers replace most physical shopping with online shopping? Two apparently competing hypotheses were tested: the diffusion of innovation hypothesis, suggesting openness to new technologies; and the efficiency hypothesis, suggesting accessibility gains. Whether the innovation hypothesis has lost its importance in favor of the efficiency hypothesis was also questioned. The study area was a polycentric urban area in the Netherlands. We distinguished between books, clothes, and groceries. It was assumed that shoppers' decisions to buy a particular good online or not, and the share of online shopping relative to in-store shopping for this good, were basically driven either by shoppers' willingness to adopt the new technology of e-shopping or by shoppers' accessibility to shops. Support was found for both hypotheses, although the impact of shopping seemed limited and varied between different types of goods. In the end, e-shopping behavior remains primarily shaped by households who are open to new technologies, and to a limited extent by efficiency considerations.

Online shopping has grown dramatically in the past decade for almost all types of goods, varying from books to clothes and groceries. As of 2017, the share of the population buying online is 83% in China, about 80% for countries in Western Europe, and 77% for the United States (*1*, *2*).

Without doubt, the increasing penetration of e-retail will have major effects on physical stores (*3*). Shopping is one of the most important non-work activities in any affluent society, in relation to money and time. With the rising level of e-shopping, the level of store shopping activities will decline, and consequently the amount of shopping floor space. It is unclear however, what the consequences will be for the spatial distribution of retail services.

Although the distribution of physical shopping areas varies between countries, the emergence of e-shopping may well reduce these differences. U.S. shopping areas are generally concentrated in malls at the fringes of the urban areas, highly, and often exclusively, accessible by car. Nevertheless, there is also some opposition from the Smart Growth movement, which aims for mixed-use development. On the contrary, in Europe, it is mainly the city centers that offer a wide variety of consumer goods. Although they also provide daily goods, the majority of supermarkets, bakers, groceries, and drug stores are found in neighborhood shopping centers. However, partly driven by the scale-up of the retail sector, many European cities are showing a tendency toward the spatial model that is common in the United States, and consequently a decline in shopping facilities in city centers, residential neighborhoods, and villages (*4*–*6*). Early examples are the French hypermarchés. In the Netherlands, the concentration of shopping facilities in urban areas has always been preserved, forced by policies restricting non-urban shopping areas, to limit travel distances and enable households to use alternatives to the car (*7*). Since the early 2000s, however, policies have been relaxed, which has now also led to a trend toward shopping facilities being located in the city fringes. Moreover, in recent years, many countries have experienced a drop in support for traditional village shops, often resulting in widespread closures.

If, as a result of increasing e-shopping, the loss of physical stores occurs mainly in urban centers then that will erode the amenities function that they currently perform. The disappearance of shops from the suburbs and rural villages has been particularly unfavorable because of a loss in variety of shops, accessibility of the shopping facilities, and their social function.

[1]Faculty of Architecture and The Built Environment, Delft University of Technology, Delft, the Netherlands

Corresponding Author:
Address correspondence to Kees Maat: c.maat@tudelft.nl

The question is: In which areas will consumers replace most physical shopping with online shopping? As early as 2003, Anderson et al. formulated two apparently competing hypotheses (*8*); Cao et al. adopted these and referred to them as the diffusion of innovation hypothesis and the efficiency hypothesis (*9*). The first hypothesis argues that urban areas facilitate innovation and creativity. People here are more open to early adoption of new technologies such as e-shopping, partly because they are often better educated and more affluent. The second hypothesis claims that one of the advantages of e-shopping is that it reduces the spatial constraint of low retail accessibility. People on the periphery therefore benefit most from the online accessibility of goods and will therefore adopt e-shopping. An intriguing prediction that Anderson et al. make is that the innovation hypothesis will mainly take place in the short run (*8*). In the longer run, when technology has been diffused, accessibility will become the main driver.

Apart from these drivers for online shopping, the decision to buy a particular good online or not will also depend on the type of good. We distinguished between search goods and experience goods (*8, 10, 11*). In the first category, required information can be offered more efficiently on the Internet, such as material on books and music. This is also the category that has been offered online for a long time. The second category, on the other hand, consists of goods that buyers want to see and try, and which are therefore more difficult to sell online, such as clothing and groceries.

As Cao et al. have already noted, little research has been done into the spatial distribution of e-shopping and the two hypotheses have hardly been tested empirically, which calls for more research to test generalizability (*9*). In this paper we also question whether, now that e-shopping seems to have made a breakthrough, the innovation hypothesis has lost its importance in favor of the efficiency hypothesis. The study area is a polycentric urban area in the Netherlands. We distinguish between books, clothes, and groceries. It is assumed that shoppers' decisions (*i*) whether to buy online or not for a particular good, and (*ii*) in relation to the share of online shopping relative to in-store shopping for this good, are basically driven by two processes. Therefore, we estimate a so-called zero-inflated model. In the next sections, we first describe the literature in this field, then we describe the data and the methods, followed by the model results, and end with conclusions.

Literature

In the past two decades, a large number of studies have been carried out in which online and physical shopping were related to the location. Weltevreden and Rietbergen show that city centers are more likely to face a reduction in shopping visits and purchases due to online shopping than rural shopping centers (*12*). The higher vulnerability of city centers is to some extent surprising in view of their larger number and variety of shops, and also their more appealing atmosphere/environment and therefore ability to attract (recreational) shoppers. On the other hand, the limited effect on rural shopping centers could be explained by their more important function for daily goods purchases (e.g., groceries) that are less sensitive to online shopping.

Urban consumers have been found to shop online more than suburban and rural residents (*13–16*). Farag et al. found that 61% of online shoppers in the Netherlands in 2001 lived in very strongly urbanized areas, whereas 39% resided in less urbanized and non-urbanized areas (32% and 7%, respectively) (*14*).

These findings seem contrary to expectations regarding the influence of store accessibility on online shopping. Since convenience, saving time, and saving money are the most frequently cited reasons for online shopping these motivational factors are likely to be less strong for urban residents that have relatively better access to shops compared with their rural counterparts. When these rural residents visit shops they need to travel over longer distances, particularly when they need a variety of products (chaining of shopping trips) causing more travel time and higher travel costs. Furthermore, due to the larger distances to the shops, these rural residents may be dependent on other transport modes, for example the car, than their urban counterparts. And car use to access shops may bring additional inconveniences, such as finding a parking place, and costs, for example, parking fees. It is, however, interesting to note that, although Farag et al. (*14*) discerned a positive relation between store accessibility and online buying they also found that better access to shops also leads to more shopping trips made (probably to explore a product in-store, since a negative relation was found between store accessibility and online search frequency). Thus, Farag et al. postulate that persons who like to shop in-store probably also like to shop online. All in all, the studies' conclusions regarding the role of accessibility are rather uncertain.

Other studies that tried to identify a relationship between store accessibility and online shopping behavior had mixed findings. Consistent with Farag et al., Cao et al. found store accessibility was positively associated with online shopping in urban areas (*9, 14*). However, for suburban areas no effect of store accessibility was found, whereas the impact in the exurban areas was negative, i.e., lower store accessibility induces more online shopping (*9, 14*). Ren and Kwan similarly found a negative effect on online shopping from store accessibility, but the measure effect was small (*15*). On the other hand,

Weltevreden and Rietbergen came to other conclusions regarding the relevance of store accessibility in city centers (*13*). They examined the role of perceived city center attractiveness by distinguishing three factors: shopping attractiveness, leisure attractiveness, and accessibility attractiveness, which was measured by the total number of shops a person can reach by car, bicycle, or on foot from his/her place of residence in a certain time distance, and taking into account the available parking for cars/bicycles/mopeds and parking fees for cars. The authors found evidence that the higher the perceived city-center attractiveness the lower the probability that consumers make fewer purchases at city-center stores due to online shopping. However, the importance of accessibility attractiveness differed among transport users. With regard to car users, the perceived car accessibility of the city center influenced the extent to which they shopped online, and the extent to which they purchased less in city centers owing to e-shopping. For users of other transportation modes, perceived shopping attractiveness was more important.

Studies in which only the nearest distance to shops is used to measure store accessibility showed different findings. Sinai and Waldvogel found that the further people live from the nearest book or clothing store, the more books or clothing they relatively buy online (*17*). In contrast, Lee et al. reported that online shopping is not affected by distance to the nearest shopping center (*18*).

Studies on the impact of spatial and accessibility aspects have always controlled for personal, attitudinal, and household aspects. Without exception, these studies show that such factors have a larger impact than accessibility. Although we assume the role of such factors is diminishing, it is still necessary to control for them. As research on online shopping has been dominated by marketing research, shopping attitudes have been extensively investigated. Research has generally focused on people's motives for shopping (e.g., acquire goods, socialize, entertain) and the relevance of the particularities of in-store shopping versus online shopping (shopping mode characteristics). Motivational factors of individuals (e.g., functional or recreational shopping) play a key role in the decision whether to buy online or in-store (*14*, *19*). Recreational shoppers are more inclined to like seeing and handling products before buying and hence will make fewer online purchases, whereas functional shoppers like to save time spent on shopping and hence buy online more frequently (*19–21*). In relation to time saving the flexibility of opening hours is often quoted as a condition that encourages online shopping (*22*). On the other hand, studies have addressed some disadvantages of the online shopping mode, that is, the inconvenience of returning products and uncertainty about the security of transactions (*23*). However, since electronic payments have boomed it is unlikely that today the latter is still a concern and mitigating factor for online shopping. Nowadays privacy concerns seem to be a more serious issue (*24*, *25*). Other behavioral studies have pointed to the relevance of a positive attitude toward online shopping, such as the perceived quality of vendors on the Internet, on the intention to buy online (*27*).

Various studies have found a positive relationship between Internet usage and online shopping, whereas other studies have not (*26*, *28*). Although more recent studies also found the Internet experience to have a positive influence on online shopping, it may well be that the effects of the Internet experience are diminishing because of the increasing penetration of Internet in our daily life (*24*, *29*). As regards the influence of mobile technology (smart phones and tablets) on online shopping (M-shopping) there is very little known yet. M-shopping is in fact still in its infancy, but is expected to gain popularity due to the increased use of smartphones. Most studies undertaken so far have addressed issues regarding its adoption by consumers. Among the few studies that focused on the impact of M-shopping on purchases, Wang et al. found an increase in purchases (*30*). However, using mobile devices can be supportive for in-store shopping, for instance in finding directions, store locations, and opening hours.

Product characteristics also affect online shopping. Search goods, such as books and CDs, are more suited to purchasing via the Internet than experience goods such as fresh vegetables (*14*, *22*, *29*, *31*). Moreover, people tend to buy more expensive products online because of possible larger savings compared with prices in physical shops (*30*).

Socioeconomic and demographic characteristics are among the most frequently studied factors in online shopping research (*14*, *23*, *29*, *31–35*). In the early stages of online shopping, male shoppers made more online purchases and spent more money online than females, but women have caught up. More recent studies have found that gender has little to no effect on online shopping activity (*18*, *36*, *37*). However, men and women are rather different in their preference for products online, with women more often purchasing clothing and men more likely to shop for electronics (*38*).

The online-shopper profile is dominated by young people, as they are typically more likely to adopt new technologies, live in households with Internet access, and have higher levels of Internet use. However, a significant increase in older (retired) online shoppers can be expected due to an aging society and the increasing engagement of this age group in information and communications technology applications such as online shopping, virtual communities, and online learning (*37*). Ultimately, online shopping might be a welcome solution for less-mobile older shoppers.

Online shoppers tend to be in the higher income groups than traditional in-store shoppers, most likely because demand increases as income increases. This could point to online shopping as a complimentary activity.

According to several studies, online shoppers are more educated (e.g., *18*, *19*, *26*, *34*), but there are also studies that found no effect of education on online shopping (e.g., *35*). The latter possibly supports the notion that online shopping is actually a relatively easy task that does not require higher education. Households with children tend to engage more often in shopping than households without children, but there is little evidence for any impact of this difference in household type on online shopping. As households with children are expected to be more time-pressured, it is likely that such households are more inclined to shop online. Some support for this was found by Hiselius et al. who observed a higher share of online shoppers among families with children under 10 years of age (*40*). Farag et al. discerned that singles shop less often online than other household types (*20*). However, it is very doubtful that their explanation that singles often do not have a fast Internet connection is still valid today.

As far as other socioeconomic aspects are concerned, car ownership or access to a car and having a driver's license tend to be negatively related to online shopping (*28*). A possible explanation for this is that the possibility to take goods from the store back home by car makes in-store shopping less inconvenient for people than those who do not have access to a car (*41*).

Methodology

Study Area

Data was collected in the Leiden urban region in the Randstad area in the Netherlands. The Randstad is the economic and population core of the Netherlands. As our research question is spatially oriented, and therefore we selected a medium-sized polycentric urban region with a core city, suburbs and exurban villages. The city of Leiden is such a medium-sized polycentric urban region, with a historical city center, surrounded by city neighborhoods, suburbs and rural villages. Leiden's city center is the main shopping area for nondaily goods, with an emphasis on clothes. Leiderdorp is a suburb, adjacent to Leiden, with a highly educated population and high income groups. Merenwijk is a medium-income neighborhood from the 1970s with mainly single family homes. Tuinstad is a neighborhood that was created between the 1930s and 1950s, close to the city center. Fortuinwijk is a 1960s neighborhood with mainly cheap apartment flats. Warmond is a larger village, Hoogmade and Zoeterwoude are smaller and relatively remote villages. Some neighborhoods have their own shopping center with the focus on daily goods (such as food and personal care); the villages have only a limited number of shops.

Data

A personal survey was conducted in 2014, in the first week of December. In each of the residential areas, respondents were selected and asked to complete a short questionnaire at their home. They were randomly selected on a map to ensure a good spatial distribution across the neighborhoods. If nobody was at home, the surveyors went to the neighbors. The questionnaire took them at most 10 min. The address has been added by the surveyors. The total sample of (almost) complete questionnaires consisted of 534 respondents. For households that were at home, the response rate was about 20%; not at home was not recorded. Thanks to this face-to-face method, a reliable representation over the selected neighborhoods was ensured.

The sample reflects the profile of the Leiden urban area. Females are slightly over-represented, although this varies over the neighborhoods. In general the respondents are well-paid workers, predominantly belonging to medium-income classes. The educational level is over-represented compared to Dutch averages, which makes sense as Leiden is a university town with science-related business clusters, and also because higher educated people are usually more willing to participate in surveys.

In addition to gender, income and education the following demographic and socioeconomic characteristics that may influence respondents' e-shopping behavior have been included in the survey: age, employment status, household composition (limited to children or not) and car availability. To discover the possible impact of the affinity with Internet use on e-shopping the survey included a question on time that is spent on Internet use and use of smartphone.

The three product categories focused on in the current research were books, a so-called search good, and one of the first shopping goods to appear online; clothes is the fastest growing category, an experience good, which people want to see and try on; and groceries, also a "see and feel" category, referred to as daily shopping. For each of these product groups, respondents were asked how often they purchased these products in-store and online. Responses were collated on a ratio scale, that is, the number of purchases per month or year. A seven-point Likert scale was used to measure participant responses to gain insight into attitudes toward shopping for these three product groups and toward online shopping in general. The possible answers ranged from "very unpleasant" to "very pleasant."

Table 1. Sample Descriptives

Residential location	Leiden city center	Warmond	Merenwijk	Tuinstad	Fortuinwijk	Hoogmade/ Zoeterwoude	Leiderdorp	Total
Share, %								
Female	56.6	59.7	64.1	61.2	66.7	68.2	50.0	60.9
Income								
Low	34.2	32.5	21.8	16.3	18.0	20.5	11.4	22.1
Medium	51.3	53.3	65.4	57.1	76.9	69.3	50.0	60.7
High	14.5	14.3	12.8	26.5	5.1	10.2	38.6	17.2
Education								
Low	7.9	32.5	11.5	2.0	24.4	21.6	6.8	15.9
Medium	23.7	32.5	23.1	30.6	26.9	42.1	22.7	28.8
High	68.4	35.1	65.4	67.4	48.7	36.4	70.5	55.2
Children	26.3	40.3	57.7	51.0	34.6	59.1	70.5	49.1
Car available	47.4	83.1	76.9	79.6	79.5	89.8	93.2	79.0
Smartphone available	80.3	59.7	78.2	85.7	68.0	77.3	90.9	77.0
Daily internet user	90.8	72.7	91.0	93.9	85.9	85.2	96.6	87.8
Average								
Age	39.1	53.9	48.5	41.5	49.8	43.6	45.5	46.2
Accessibility								
Potential clothes	1.8	0.9	1.0	1.2	1.0	0.9	0.9	1.1
Potential books	0.4	0.2	0.3	0.3	0.3	0.2	0.2	0.3
Potential groceries	4.8	3.7	4.1	4.5	4.3	3.7	3.9	4.1
Walkscore	94.9	71.1	62.0	82.9	64.8	46.5	45.7	65.1
Sample frequency	76	77	78	49	78	88	88	534

Note: Income and education categories add up to 100%; the other shares are binaries, so they add up to 100% with their counterpart.

Furthermore, to investigate the impact of general attitudes to e-shopping behavior the respondents were asked to provide some general opinions. The questions included attitudes toward family, technology, environment, culture, sports, and employment. Respondents could answer each question on a seven-point Likert scale, ranging from "very unimportant" to "very important."

Shopping accessibility measures were developed by relating the zip code of the respondents to data about the location of shops and their number of retail jobs in the Leiden region. Three different shopping accessibility indicators were created. A potential accessibility indicator (or gravity-based indicator) was calculated on the basis of the number of shops, weighted by home-shop distances and a distance decay (i.e., shops at further distance get a lower weight in the calculation). A second accessibility indicator was calculated by summing the number of shops within 5-km buffers. The third indicator, Walkscore (www.walkscore.com), was based on shopping in the direct neighborhood, using Google Maps data, with a distance decay also applied. Walkscore is the most local indicator of all and has a restricted upper distance value. It has the largest potential distance range (basically to infinity, although strongly limited by the distance decay). All three indicators were specified and applied in our model estimations, but only the potential accessibility indicator led to significant estimates. Therefore the results concern only the models in which the potential accessibility indicator was used. The values of the potential accessibility indicator as well as other important descriptive statistics of the sample are presented in Table 1.

Modeling Approach

We estimated models for each of the three categories of goods. For each category, we asked about both the number of in-store and online purchases per month or per year and recalculated this to the number of purchases per month. Then we calculated for each goods category the share of online purchases in the total number of purchases. An appropriate model for shares is a fractional regression model, which exists both in a logit and a probit version (*42*). Fractional responses concern outcomes between zero and one, for example the average of dichotomous variables such as participation rates, or variables that are naturally on a zero to one scale, such as pollution levels. Or, as in the present study, the number of online shopping purchases as a share of the total purchases (for a particular good). If we assume that almost all people shop both online and in-store, and only some of them just do online or in-store shopping, we could simply calculate shares, which would vary from zero to one: zero for those who just shop in-store and a one for those who just shop online, and all others in between. In that case, we would simply assume that such shares

Table 2. Models for Online/In-Store Clothes Shopping

Binary logit models	(a) Including accessibility potential		(b) Including neighborhoods	
	Odds ratio	*p*-value	Odds ratio	*p*-value
Store accessibility: potential (z-value)	1.422	.022		
Residential location				
Warmond			.266	.002
Merenwijk			.418	.044
Tuinstad			.375	.035
Fortuinwijk			.264	.002
Hoogmade/Zoeterwoude			.401	.035
Leiderdorp			.468	.087
Leiden city center (ref)				
Smartphone	1.751	.056	1.653	.088
Daily Internet use	3.428	.001	3.317	.002
Female (male = ref)	2.495	.000	2.611	.000
Age	.974	.000	.974	.001
Children (no children = ref)	1.662	.024	1.597	.041
Household income (low = ref)				
Medium	1.541	.114	1.545	.119
High	2.040	.048	1.916	.079
Technology interest	1.153	.057	1.162	.055
Culture interest	.797	.006	.796	.007

Note: (a) R-sq McFadden = .186; Cragg/Uhler = .296; N = 514; (b) R-sq McFadden = .195; Cragg/Uhler = .309; N = 513.

represent the same decision-making process, and would be influenced in the same way by a certain set of predictors. In such cases, a fractional logit model is the most appropriate model. However, if we observed that a large share of the respondents just bought in-store, that would mean the data was zero-inflated. Therefore, we assumed that shoppers' decisions whether to buy online or not for a particular good, and the share of online purchases, are two distinct processes. Hence, we developed two model types. For the decision to be involved in online shopping for a particular good, we developed a binary logit model, whereas we estimated a fractional logit model for the fraction above zero. Basically, it would make sense to not only estimate models taking into account zero-inflation, but also one-inflation, assuming that a person only buys online. However, as this is (still) rather rare, we did not estimate this. An alternative model for the fractional logit model is the beta regression model, but this model type cannot deal with zeros and ones at all. Model fit is approached by pseudo R-squares (rho); in binary logit it is assumed that Cragg and Uhler better approaches the proportion of explained variance.

Results

Clothes

Table 2 shows two models estimating the likelihood that someone shops for clothes online. The left model (a) estimates the likelihood that someone shops online with the potential an indicator for accessibility. Model (b) is comparable with model (a) but uses the residential location with Leiden city center as the reference category. The two models are more or less comparable, although the pseudo R-squares suggest that the residential location model performs slightly better.

Model (a) suggests that the more accessible the clothing shops (i.e., an increasing potential value), the greater the share of people shopping online. At first sight, this sounds counterintuitive, as people with a high accessibility have more in-store shops at their disposal and have, consequently, less need to shop online. To get a better insight, we inspected the model in which potential was replaced with residential location. We then found that respondents in the city center were most likely to shop online, and the inhabitants of the less affluent neighborhood of Fortuinwijk and the village of Warmond were least likely. This suggests that there are other mechanisms at play, and confirms the innovation hypothesis. For completeness, we also tested simple buffers and Walkscore, however, they were not significant.

The impact of the other variables was as expected: those who use the Internet on a daily basis and also those who have a smartphone (rather than a simple phone or no mobile phone at all), significantly shop more online. Previous literature found that more men shop online, however this seems an effect of the early days, as now women shop online much more often, especially for clothes. We found households with children had a greater likelihood of shopping online. Age was negatively correlated with online shopping. Educational level and

Table 3. Models for Online/In-Store Clothes Shopping

	(a) Including accessibility potential		(b) Including neighborhoods	
Fractional logit models	Odds ratio	*p*-value	Odds ratio	*p*-value
Store accessibility: potential (z-value)	0.898	.036		
Fun shopper	0.906	.002		
Residential location				
Warmond			1.688	0.006
Merenwijk			1.735	0.004
Tuinstad			1.505	0.042
Fortuinwijk			1.342	0.107
Hoogmade/Zoeterwoude			1.300	0.111
Leiderdorp			1.326	0.077
Leiden city center (ref)				

Table 4. Model for Online/In-Store Grocery Shopping

Binary logit model	Odds ratio	*p*-value
Household income (low = ref)		
Medium	1.81	0.092
High	2.10	0.074

Note: R-sq McFadden = .009; Cragg/Uhler = .013; *N* = 514.

household income were almost comparable, but income was included as it explains the likelihood of shopping online slightly better. We also tested the effects of being a student or having a job, but these were not significant.

Next we estimated the share of online shopping for those who shop for clothes online (Table 3) using fractional logit. We found a negative relation between shopping accessibility and the proportion of online shopping. In addition, an interesting observation was that the negative effect of the respondents' attitude toward "fun" shopping suggests that people who enjoy shopping tend to purchase relatively more in-store.

Groceries

Shopping online for groceries is still in its infancy in the Netherlands. In our sample only 81 respondents (15%) bought groceries online. Groceries are typically experience goods. The role of shopping accessibility may also differ due to the need to purchase groceries frequently. Therefore store accessibility can be assumed to be important and hence, we can hypothesize that lower accessibility will go together with more online shopping.

The results of the binary logit analysis show, however, that shopping accessibility does not affect the likelihood of online grocery shopping. The only significant effect is income (Table 4). Higher income households are more inclined to buy groceries online. This confirms the innovation hypothesis, and is also plausible as home delivery for groceries comes with a delivery fee, to which higher income households are less sensitive. It is not surprising that no significant differences were found for accessibility, as grocery shops in the Netherlands are almost always nearby. The estimates of the fractional model did not lead to any significant effects.

Table 5. Models for Online/In-Store Book Shopping

Binary logit model	Odds ratio	*p*-value
Store accessibility: potential (z-value)	1.29	0.059
Daily Internet use	4.37	0.001
Education (low = ref)	2.10	0.074
Medium	1.45	0.394
High	2.55	0.023
Pleasure in e-shopping	1.39	0.001
Fractional logit model	**Odds ratio**	***p*-value**
Culture interest	0.88	0.007
Pleasure in shopping books	0.84	0.001

Note: For binary logit model, R-sq McFadden = 0.009, R-sq Cragg-Uhler = 0.013. For fractional logit model, R-sq McFadden = 0.024.

Books

Books are search goods, which makes them very suitable to buy on the Internet. We found that higher shopping accessibility increases the likelihood of online shopping. This finding is contrary to the efficiency hypothesis; possibly the presence of book shops nearby and likely visits to these shops inspire readers to buy books, which they actually do not buy in-store but online. Buyers of books that are higher educated, spend more time on the Internet, and enjoy shopping online in general are also more inclined to buy online (Table 5).

The results of the fractional model demonstrate that the proportion of online shopping for books is not related to shopping accessibility. So there is a higher probability that consumers will buy books online when book stores are nearby, but the proximity of these stores will not have an impact on the proportion of books bought online. On the other hand, interest in culture and pleasure in shopping for books do have an impact. It appears that buyers of books that have an interest in culture and take pleasure in buying books purchase less online than their counterparts.

Conclusions

The increasing penetration of e-retail will have major effects on physical stores, and the question is: In which areas will consumers replace most physical shopping with online shopping? Two apparently competing hypotheses were tested: the diffusion of innovation hypothesis and the efficiency hypothesis, formulated by Anderson et al. (*8*). In this paper we also tested these hypotheses by analyzing the influence of shopping accessibility on online shopping, controlling for demographic and socioeconomic factors and shopping-related attitudes. For this purpose two models were developed to explicitly consider the shoppers' decisions regarding (i) whether or not to buy online, and (ii) the share of online shopping relative to in-store shopping. The models were estimated for three categories of goods (clothes, groceries, and books).

We found support for both hypotheses. For clothes shopping the results suggest that people with greater access to shops are more likely to buy on the Internet, which did not confirm the hypothesis of efficiency in relation to accessibility. However, a more likely explanation could be that people with greater shopping accessibility, who live in the city center more often, have a different demographic and socioeconomic profile (younger, higher educated, Internet-oriented) from their counterparts that makes them more inclined to buy online. This therefore confirms the innovation hypothesis. On the other hand, analyzing the proportion of online shopping compared with store shopping revealed that online consumers with lower access to shops purchase more online. This is plausible, considering that these consumers face greater inconvenience (time and cost) when visiting shops and therefore may use online shopping to reduce the number of trips. This supports the efficiency hypothesis.

For grocery shopping our study did not find any direct evidence for the influence of shopping accessibility on shopping online. In general, the inconvenience of visiting stores is low in the Netherlands, because of a high density of grocery shops. Moreover, the number of consumers that buy groceries online is small as it is still a rather new shopping service that is also not yet widely offered. Our findings demonstrate that higher income households are inclined to use online grocery shopping, as they are less sensitive to the additional costs of home delivery, and may be also more open to new technologies, as the innovation hypothesis suggests.

For book shopping we had a somewhat surprising result regarding the impact of accessibility on online shopping. Potential buyers living near bookstores are more likely to buy online than their counterparts. It seems that the presence of bookstores nearby and likely visits to these stores inspire people to buy books that are actually not bought in these in-store shops but online. The motivation for this behavior is, however, unclear. Conversely, we found that taking pleasure in buying books and an interest in culture discourage online shopping and encourage buyers to purchase books in-store.

Overall, we can conclude that shopping accessibility may have an impact on online shopping, but its impact seems limited and varies between different types of goods. In the end, e-shopping behavior remains primarily shaped by households who are open to new technologies, and only to a limited extent by efficiency considerations.

Author Contributions

The authors confirm contribution to the paper as follows: study conception and design, data collection, analysis and interpretation of results, draft manuscript preparation: KM, RK. Both authors reviewed the results and approved the final version of the manuscript.

References

1. *Worldwide Online Retail Rate 2017, by Country; Global Markets with the Highest Online Shopping Penetration Rate as of 2nd Quarter*. Statista, Hamburg, Germany, 2017. www.statista.com/statistics/274251/retail-site-penetration-across-markets. Accessed March 1, 2018.
2. Online winkelen; kenmerken aankoop, persoonskenmerken. CBS, Voorburg/Heerlen. http://statline.cbs.nl/Statweb. Accessed March 1, 2018.
3. Salomon, I. Telecommunications and Travel Relationships: A Review. *Transportation Research Part A: General*, Vol. 20. No. 3, 1986, pp. 223–238.
4. Guy, C. M. Off-centre Retailing in the UK: Prospects for the Future and the Implications for Town Centres. *Built Environment*, Vol. 24, No. 1, 1998, pp. 16–30.
5. Guy, C. M. Controlling New Retail Spaces: The Impress of Planning Policies in Western Europe, *Urban Studies*, Vol. 35, No. 5-6, 1998, pp. 953–979.
6. Thomas, C. J., R. D. F. Bromley, and A. R. Tallon. Retail Parks Revisited: A Growing Competitive Threat to Traditional Shopping Centres? *Environment and Planning A*, Vol. 36, 2004, pp. 647–666.
7. Evers, D. The Rise (and fall?) of National Retail Planning. *Tijdschrift voor Economische en Sociale Geografie*, Vol. 93, No. 1, 2002, pp. 107–113.

8. Anderson, W. P., L. Chatterjee, and T. R. Lakshmanan. E-commerce, Transportation, and Economic Geography. *Growth and Change*, Vol. 34, No. 4, 2003, pp. 415–432.
9. Cao, X. J., Q. Cehn, and S. Choo. Geographic Distribution of E-shopping; Application of Structural Equation Models in the Twin Cities of Minnesota. *Transportation Research Record: Journal of the Transportation Research Board*, 2013. 2383: 18–26.
10. Zhai, Q., X. Cao, P. L. Mokhtarian, and F. Zhen. The Interactions Between E-Shopping and Store Shopping in the Shopping Process for Search Goods and Experience Goods. *Transportation*, Vol. 44, No. 5, 2017, pp. 885–904.
11. Cao, X. The Relationships Between E-Shopping and Store Shopping in the Shopping Process of Search Goods. *Transportation Research Part A*, Vol. 46, No. 7, 2012, pp. 993–1002.
12. Weltevreden, J. W. J., and T. van Rietbergen. The Implications of E-Shopping for In-Store Shopping at Various Shopping Locations in the Netherlands. *Environment and Planning B*, Vol. 36. 2009, pp. 279–299.
13. Weltevreden, J. W. J., and T. van Rietbergen. E-shopping Versus City Centre Shopping: The Role of Perceived City Centre Attractiveness. *Tijdschrift voor Economische en Sociale Geografie*, Vol. 98, No. 1, 2007, pp. 68–85.
14. Farag, S., J. Weltevreden, T. van Rietbergen, M. Dijst, and F. van Oort. E-shopping in the Netherlands: Does Geography Matter? *Environment and Planning B: Planning and Design*, 33, 2006, pp. 59–74.
15. Ren, F., and M. P. Kwan. The Impact of Geographic Context on E-Shopping Behavior. *Environment and Planning B: Planning and Design*, Vol. 36, 2009, pp. 262–278.
16. Sener, I. N., and P. R. Reeder. An Examination of Behavioral Linkages Across ICT Choice Dimensions: Copula Modeling of Telecommuting and Teleshopping Choice Behavior. *Environment and Planning A*, Vol. 44, No. 6, 2012, pp. 1459–1478.
17. Sinai, T., and J. Waldfogel. Geography and the Internet: Is the Internet a Substitute or a Complement for Cities? *Journal of Urban Economics*, Vol. 56, No. 1, 2004, pp. 1–24.
18. Lee, R. J., I. N. Sener, and S. L. Handy. A Picture of Online Shoppers: Specific Focus on Davis, California. *Transportation Research Record: Journal of the Transportation Research Board*, 2015. 2496: 55–63.
19. Li, H., C. Kuo, and M. G. Russell. The Impact of Perceived Channel Utilities, Shopping Orientations, and Demographics on the Consumer's Online Buying Behaviour. *Journal of Computer Mediated Communication*, Vol. 5, No. 2, 1999. http://www.ascusc.org/jcmc/vol5/issue2/hairong.html.
20. Farag, S., T. Schwanen, M. Dijst, and J. Faber. Shopping Online and/or In-Store? A Structural Equation Model of the Relationships Between E-Shopping and In-Store Shopping. *Transportation Research Part A*, Vol. 41, 2007, pp. 125–141.
21. Verhoef, P. C., and F. Langerak. Possible Determinants of Consumers' Adoption of Electronic Grocery Shopping in the Netherlands. *Journal of Retailing and Consumer Services*, Vol. 8, 2001, pp. 275–285.
22. Lee, P. Behavioural Model of Online Purchasers in E-commerce Environment. *Electronic Commerce Research*, Vol. 2, 2002, pp. 75–85.
23. Forsythe, S., and B. Shi. Consumer Patronage and Risk Perceptions in Internet Shopping. *Journal of Business Research*, Vol. 56, 2003, pp. 867–875.
24. Akther, S. H. Who Spends More Online? The Influence of Time, Usage Variety, and Privacy Concern on Online Spending. *Journal of Retailing and Consumer Services*, Vol. 19, No. 1, 2012, pp. 109–115.
25. Pain, C., U.-D. Reips, S. Stieger, A. Joinson, and T. Buchanan. Internet Users' Perceptions of 'Privacy Concerns' and 'Privacy Actions.' *International Journal of Human-Computer Studies*, Vol. 65, No. 6, 2007, pp. 526–536.
26. Liao, Z., and M. T. Cheung. Internet-based E-shopping and Consumer Attitudes: An Empirical Study. *Information & Management*, Vol. 38, No. 5, 2001, pp. 299–306.
27. Sim, L. L., and S. M. Koi. Singapore's Internet Shoppers and their Impact on Traditional Shopping Patterns. *Journal of Retailing and Consumer Services*, Vol. 9, No. 2, 2002, pp. 115–124.
28. Cho, J. Likelihood to Abort an Online Transaction: Influences from Cognitive Evaluations, Attitudes and Behavioral Variables. *Information & Management*, Vol. 41, 2004, pp. 827–838.
29. Nysveen, H., and P. Pedersen. An Exploratory Study of Customers' Perception of Company Websites Offering Various Interactive Applications: Moderating Effects of Customers' Internet Experience. *Decision Support Systems*, Vol. 37. No. 1, 2004, pp. 137–150.
30. Crocco, F., L. Eboli, and G. Mazzulla. Individual Attitudes and Shopping Mode Characteristics Affecting the Use of E-Shopping and Related Travel. *Transport and Telecommunication*, Vol. 14, No. 1, 2013, pp. 45–56.
31. Wang, R. J. H., E. C. Malthouse, and K. Lakshman. On the Go: How Mobile Shopping Affects Customer Purchase Behavior. *Journal of Retailing*, Vol. 9, No. 2, 2015, pp. 217–234.
32. Vrechopoulos, A. P., G. J. Siomkos, and G. I. Doukidis. Internet Shopping Adoption by Greek Consumers. *European Journal of Innovation Management*, Vol. 3, 2001, pp. 142–152.
33. Swinyard, W. R., and S. M. Smith. Why People (don't) Shop Online: A Lifestyle Study of the Internet Consumer. *Psychology & Marketing*, 2003, pp. 567–597.
34. Susskind, A. Electronic Commerce and World Wide Web Apprehensiveness: An Examination of Consumers' Perceptions of the World Wide Web. *Journal of Computer-Mediated Communication*, Vol. 9, No. 3, 2004.
35. Mahmood, M. A., K. Bagchi, and T. Ford. On-Line Shopping Behavior: Cross-Country Empirical Research. *International Journal of Electronic Commerce*, Vol. 9, No. 1, 2004, pp. 9–30.
36. Zhou, L., L. Dai, and D. Zhang. Online Shopping Acceptance Model – A Critical Survey of Consumer Factors in Online Shopping. *Journal of Electronic Commerce Research*, Vol. 8. No. 1, 2007, pp. 41–62.
37. Hernández, B., J. Jiménez, and M. J. Martín. Age, Gender and Income: Do They Really Moderate Online Shopping

Behaviour? *Online Information Review*, Vol. 35, No. 1, 2011, pp. 113–133.

38. Lian, J.-W., and D. C. Yen. Online Shopping Drivers and Barriers for Older Adults: Age and Gender Differences. *Computers in Human Behavior*, Vol. 37, 2014, pp. 133–143.
39. Passyn, K. A., M. Diriker, and R. B. Settle. Images of Online Versus Store Shopping: Have the Attitudes of Men and Women, Young and Old Really Changed? *Journal of Business & Economics Research*, Vol. 9, No. 1, 2011, pp. 99–110.
40. Hiselius, L. W., L. S. Rosqvist, and E. Adell. Travel Behaviour of Online Shoppers in Sweden. *Transport and Telecommunication*, Vol. 16, No. 1, 2015, pp. 21–30.
41. Reutter, O., and U. Reutter. *Autofreies Leben in der Stadt. Autofreie Stadtquartiere im Bestand.* Verkehr Spezial Bd. 2. Vertrieb für Bau- und Planungsliteratur, Dortmund, Germany, 1996.
42. Baum, C. Stata Tip 63: Modeling Proportions. *The Stata Journal*, Vol. 8, No. 2, 2008, pp. 299–303.

The Standing Committee on Effects of Information and Communication Technologies (ICT) on Travel Choices (ADB20) peer-reviewed this paper (18-00303).

Article

TRR
JOURNAL OF THE TRANSPORTATION RESEARCH BOARD

Transportation Research Record
2018, Vol. 2672(50) 11–22

DOI: 10.1177/0361198118776151
journals.sagepub.com/home/trr
SAGE

How Useful is Travel-Based Multitasking? Evidence from Commuters in Portland, Oregon

Patrick A. Singleton[1,2]

Abstract

As research on travel-based multitasking—doing other things while traveling—becomes more common, some studies also investigate the quality or value of activity participation during travel. This idea of travel usefulness is a component of the positive utility of travel concept; understanding the benefits of travel-based multitasking is important for calculating accurate economic values trading-off travel time and other variables. This study analyzed travel usefulness and its potential determinants using a 2016 survey of about 650 commuters in the Portland, Oregon, area. Ordered logit models identified factors associated with subjective assessments of the overall usefulness of activity participation while on a recent commute trip. Around 90% of walk and bicycle commuters reported useful commutes, as did about half of transit commuters and auto passengers; however, half of auto drivers viewed their commutes as wasted time. Younger travelers, those with less frequent commutes, and people who reported "doing nothing" or more passive activities (window-gazing, daydreaming) on the trip were more likely to consider their commuting time to be a waste. Traveler perceptions were more closely associated with travel usefulness than sociodemographic characteristics. Results suggested that exercise and the physical activity benefits of walking and bicycling may be considered a useful form of travel-based multitasking. Overall, few common and traditionally productive multitasked activities appeared to be useful. Instead, commuters may be doing things more to pass the time than to make productive use of it. These findings offer implications for understanding travel behavior interventions and the potential use of autonomous vehicles.

Traditional transportation analysis methods, including those that underlie travel demand forecasting models and tools for transportation project appraisal, assume the demand for passenger transportation is derived from the demand for conducting activities in spatially distinct locations. A corollary of this axiom is that travel time is a disutility that travelers desire to minimize. As a result, the primary user benefit of large, mobility-enhancing transportation projects is the aggregate value of travelers' marginal travel time savings. Over the past two decades, scholars have questioned the universality of these assumptions, instead suggesting and providing evidence that some travel may be motivated by factors other than reaching activity destinations and that some people may benefit from the act of traveling itself. These perspectives are known as the positive utility of travel (PUT) concept (*1*, *2*).

One major component of the PUT is travel-based multitasking: doing other activities while traveling (*3*). (A second aspect includes benefits from the experience of traveling itself, expressed by positive emotions or symbolic fulfillment from traveling.) People who do things while traveling presumably benefit from the activities. Foremost, they may be making productive use of their travel time (*4*) by doing traditional work, maintenance, or leisure activities: writing or reviewing documents, eating a meal, reading a novel, and so forth. For some, traveling (and commuting in particular) can be a time of transition (*5*), providing a buffer between home and work and allowing the traveler to mentally prepare for obligations at the destination, or a time to relax and escape from such obligations. People who use their commutes to snooze or sleep, think, daydream, or stare out the window at the passing (natural or urban) landscape may have some of

[1]Department of Civil & Environmental Engineering, Portland State University, Portland, OR
[2]Department of Civil & Environmental Engineering, Utah State University, Logan, UT

Corresponding Author:
Address correspondence to Patrick A. Singleton: patrick.singleton@usu.edu

these goals in mind. Still other activities may be less about productivity or mental health and more about making travel less onerous or more enjoyable: checking social media, playing a game, and listening to music are all activities that can reduce the disutility of traveling.

An important question arising from the study of travel-based multitasking is about travel usefulness: how much do people value the activities they conduct while traveling? This question is especially relevant considering the importance of the value of travel time savings (VTTS) for transportation project appraisal (*6*). This measure of the willingness to pay for a marginal reduction in travel time is usually derived from travelers' revealed or stated preferences when faced with trade-offs between travel time and cost, and it is an important input to the cost–benefit analysis of major mobility-enhancing infrastructure investments. If travelers value multitasked activities, then current VTTS estimates may be biased, yielding incorrect predictions of travel behavior shifts and calculations of user benefits. Work is underway to more formally consider activity participation during travel within microeconomic time use and allocation theories (*7*), and there is emerging research that suggests travel-based multitasking may indeed affect VTTS (*8*, *9*). Understanding the usefulness of travel activities is an important part of these efforts.

Research Questions

The goal of this study is to further our understanding of the travel usefulness concept and its relationship with travel-based multitasking. Specific research questions include: How useful is travel-based multitasking? How much do people value the time they spend engaged in activities while traveling? Which activities are considered most useful? What trip and traveler characteristics are associated with travel usefulness?

These questions are answered through an analysis of the results of a 2016 survey of commuters in the Portland, Oregon, metropolitan area. The paper is structured as follows. First, a brief literature review summarizes travel-based multitasking, travel usefulness, and their associated factors. Next, the data and methods are described. Results of ordinal logit models of travel usefulness are then presented. Finally, the paper closes with a discussion of the results and implications of this study, including limitations and future work.

Literature Review

Travel-Based Multitasking

Research on travel-based multitasking was rare before the first decade of the 21st century (*10*). Since then, a growing number of studies have investigated the things people do while traveling as well as factors potentially associated with this activity participation (*11*). Research often focuses on public transit passengers (e.g., *12*, *13*) because they may be better able to engage in a greater number and variety of activities while traveling. However, a growing number of multimodal studies find significant differences between passive or passenger modes and those requiring more active attention (*14*–*17*). For instance, car drivers are more likely to be listening to music; reading, writing, and resting are more common among transit or car passengers; and exercise is nearly exclusive to walking and bicycling. Other trip characteristics, including travel time and trip purpose, also appear influential. Gender and especially age are among the most consistently associated sociodemographic characteristics: younger travelers do more technology-based activities, whereas older travelers may be more likely to read paper media. Some evidence suggests a link between attitudes or perceptions and travel-based multitasking (*14*), yet more research is needed (*11*).

Travel Usefulness

Despite a widening examination of potential determinants of travel-based multitasking, most studies still only measure or model activity participation itself. Recently, the purview of scholars has expanded to reflect an interest in knowing not just the activities people do while traveling but also the quality of travel time use. In response, some travel-based multitasking questionnaires have begun to ask travelers to assess the value, worth, or usefulness of a trip.

Three surveys of rail passengers in Great Britain in 2004, 2010, and 2014 using questions designed by Lyons and Jain offer a longitudinal perspective into subjective valuations of travel time use (*13*, *18*, *19*). The questionnaires asked travelers to report if they "made very worthwhile use of my time," "some use of my time," or whether "my time spent on this train [was] wasted." Between 2004 and 2010, there was a significant increase in the percentage of people making "very worthwhile" use of travel time (24–31%) and a significant decrease in reports of "wasted time" (19–14%); few changes occurred between 2010 and 2014. Commuters were less likely and business travelers more likely to view their time as worthwhile. Peak period travelers had somewhat more wasted time than rail passengers traveling off-peak. Almost twice as many adults aged 65 + had worthwhile trips than did adults aged < 35 (38% vs. 21%).

Susilo et al. (*20*) examined subjective assessments of travel time use from the 2010 Great Britain rail passenger dataset in more detail using multinomial logit models, offering a multivariate analysis of potential determinants. Engagement in traditionally productive activities—studying or working, texting or calling for work, and checking emails—appeared to be more worthwhile, whereas participation in more passive activities—window-gazing or people-watching, sleeping or snoozing—was associated with reporting wasted time. Travel time negatively affected perceived travel time usefulness for leisure travelers but not for business travelers. Among traveler characteristics, age differences were again the most substantial: younger travelers (under age 45) were more likely to report "wasted time" than older travelers.

A few multimodal studies are beginning to examine differences in perceived travel usefulness by users of different modes. In a 2011–2012 study in Northern California, Circella et al. (*17*) asked commuters to report the subjective value of travel time use on a five-point scale from "mostly wasted time" to "mostly useful time." Overall, 20% of commuters said their travel time was mostly useful, whereas only 8% reported it to be mostly wasted. Car drivers had the most wasted and least useful commutes, followed by car passengers; transit, train, and bicycle commuters had the most useful commutes. Using ordered probit models, the authors then analyzed factors associated with travel usefulness. There was a positive association between usefulness and age as well as traditional paper and electronic activity participation, but only for "passive" (car, rail, or transit passenger) modes.

Rosenfield and Zhao (*21*) also investigated the quality of travel time use across various modes, asking a slightly different question about "[making] use of your commute time effectively." Overall, car drivers reported the poorest quality of time use, whereas people walking and bicycling had higher qualities; bus and rail passengers were somewhere in the middle.

Based on the results summarized above, in general, only 10%–30% of travelers view traveling as mostly wasted time, whereas typically a greater proportion (roughly 20%–30%) report travel time to be very worthwhile or mostly useful. Age differences seem significant: younger travelers appear more likely to consider traveling to be wasted time. People doing traditional work-related activities (reading, writing, or emailing) are more likely to see travel time as being useful, and people doing more passive activities (window-gazing, people-watching) are more likely to report wasted time, although these results could vary by mode (*17*).

It is reasonable to assume that questions of travel usefulness are attempting to measure an overall assessment of the value of travel-based multitasking. However, when considering the usefulness of a trip, respondents may confound the intended benefits of multitasking (e.g., productive use of travel time, preparation for a destination activity) with enjoyment of the travel experience or the instrumental benefits of reaching a destination (*3*). This possibility should be considered when analyzing travel usefulness.

Data and Methods

This paper fits within a larger study investigating the PUT concept and its effects on mode choice (*22*). The dataset used in this study included a 30-min online questionnaire survey administered to working and commuting adults in the Portland, Oregon, region regarding their most recent commute trip from home to work. After answering detailed questions about travel-based multitasking—the subject of a separate companion paper (*16*)—respondents then answered a single summary question about travel usefulness. Participants were recruited primarily via email at workplaces between mid-October and mid-December 2016. Only 656 of the 791 people who started the survey completed enough questions to be used in these analyses. Singleton (*22*) provides more information on the data collection process.

To solicit subjective assessments of travel usefulness, the questionnaire approach to measuring travel-based multitasking was chosen. First, respondents selected from a list of 23 activities all the things they did between home and work on their most recent commute trip. Next, for each selected activity, participants reported the approximate percentage of time (0%–100% in 10% increments) they spent doing that activity during the trip. Singleton (*16*) describes how these 23 activities were selected from a master list compiled from prior studies. For this analysis, five activities were removed due to low response frequencies (< 15), and five other activities were combined into two activity groups using exploratory factor analysis; see Singleton (*16*) for details. The final activities used in this study, along with the percentage of commuters (by mode) reporting each activity, are shown in Table 1.

After soliciting responses on activity participation and (percentage) duration, the survey asked about travel usefulness: "In terms of its value to you, overall, how useful would you rate the time you spent commuting? Ignore the value of getting to your destination, and think only about the things you did while commuting and the time you spent doing them." This language was borrowed from Circella et al. (*17*). Responses were on a five-point Likert-type scale: "Mostly wasted; Somewhat wasted; Neither wasted nor useful; Somewhat useful; Mostly useful." Restrictions on the survey length prevented additional detailed questions about the quality of travel-based multitasking (*21*) from being asked.

Since the travel usefulness question was measured on an ordinal scale and responses were not normally distributed (negatively skewed and platykurtic), an ordered logit model of travel usefulness was estimated. Other studies have used multinomial logit (*20*) or ordered probit (*17*) models to predict travel usefulness. Ordered logit and ordered probit models usually yield similar results with respect to tests of parameter significance. Model

Table 1. Travel Activities and their Reported Prevalence by Commute Mode

	Commute mode				
Activity	Walk	Bicycle	Auto, driver	Auto, passenger	Transit
"Information and communications technology" (ICT)-related activities	30%	0%	9%	29%	69%
Texting, emailing, or other messaging	(30%)	(0%)	(7%)	(20%)	(56%)
Reading electronically (e-book, website)	(3%)	(0%)	(1%)	(17%)	(41%)
Using social websites or apps (Facebook, Twitter, LinkedIn, Tumblr, Instagram)	(13%)	(0%)	(2%)	(14%)	(34%)
"Passive" activities	83%	82%	49%	63%	62%
Viewing scenery; watching people	(67%)	(73%)	(29%)	(49%)	(50%)
Thinking or daydreaming	(63%)	(60%)	(37%)	(51%)	(48%)
Talking face-to-face with people you know	10%	15%	10%	71%	11%
Talking face-to-face with strangers	30%	10%	1%	6%	17%
Talking on the phone	13%	2%	11%	0%	4%
Reading print (newspaper, book, etc.)	0%	1%	0%	0%	33%
Listening to music, radio, or other audio	20%	16%	79%	66%	29%
Playing game (Pokémon Go, puzzle, etc.)	3%	0%	1%	3%	16%
Eating food; drinking beverage	17%	4%	33%	31%	14%
Personal grooming (shaving, makeup, etc.)	0%	0%	4%	14%	2%
Singing; dancing	3%	8%	8%	6%	0%
Exercising or being physically active	60%	89%	1%	0%	5%
Planning or navigating this trip	10%	11%	12%	6%	6%
Sleeping or snoozing	0%	0%	0%	11%	11%
Doing nothing	10%	6%	10%	14%	7%

estimation was conducted using the MASS (*23*) and stats packages in R.

Table 2 summarizes the independent variables used in the models and their descriptive statistics. Variables included trip characteristics (commute mode, travel time, number of co-travelers), weather (temperature, precipitation), traveler demographics and socioeconomics (individual, household, transportation, and job attributes), and traveler perceptions (satisfaction with typical commute travel time, self-reported ideal commute travel time, travel liking, and the teleportation test). These variables have been used in other travel-based multitasking studies and were available for most cases. Previous research on the PUT concept has employed the traveler perception questions (*24*, *25*). Independent variables were examined for multicollinearity issues before entering the model, and moderately-to-strongly correlated (>0.40) ones were removed.

Results

Travel Usefulness by Mode

Figure 1 displays responses to the travel usefulness questions, summarized by commute mode. Overwhelmingly, most people walking (87%) and bicycling (94%) reported having at least somewhat useful commutes, with the most useful commutes experienced by people bicycling (68% mostly useful). Slightly more than half of transit commuters (57%) and auto passengers (54%) reported useful commutes, whereas about half of auto drivers felt like their commutes were at least somewhat wasted time (50%). There appear to have been significant modal differences in reported travel usefulness.

Determinants of Travel Usefulness

To understand factors associated with ratings of travel usefulness, several ordered logit models were estimated using a variety of specifications. Model estimation results are shown in Table 3 for two different models: the primary model (A) using activity participation (yes vs. no), and an alternate model (B) using activity duration (minutes). The activity duration model is shown because it offers different interpretations of travel time and activity engagement during travel than the activity participation model. The dependent variable in both models is the ordered categorical travel usefulness variable, with "Mostly useful" as the most positive category. (The reference category for commute mode is "Auto, driver," and the reference category for travel liking is "Neither disliked nor liked.")

Because specific activities are difficult or unlikely on certain modes, and because modal characteristics likely affect the quality of activity participation, the travel activity variables were tested for significant interactions with commute mode. Each activity was interacted with a variety of possible modes or modal groupings; although, to reduce the possibility of finding spurious associations (and to ensure that models were estimable), groupings had to include a minimum 15 observations of those mode users reporting that activity. (Thus, walk/bicycle

Table 2. Descriptive Statistics

Variable	Categorical		Continuous	
	#	%	Mean	SD
Trip characteristics				
Mode: Walk	30	4.3		
Bicycle	114	16.5		
Transit	175	25.4		
Auto, passenger	35	5.1		
Auto, driver	336	48.7		
Travel time (min)			35.66	21.27
# co-travelers			0.24	0.70
Temperature (°F) Δ from average			2.71	5.15
Day precipitation ≥0.10 in	155	22.9		
Traveler sociodemographics				
Age: 18–34 years	142	19.4		
35–44 years	190	25.9		
55–64 years	174	23.7		
65 + years	48	6.5		
Gender: Female	403	55.4		
Race/ethnicity: Missing	24	3.3		
Hispanic/non-white/multiple	101	13.7		
Disability	54	7.3		
Student	54	7.3		
Education: No college degree	131	17.9		
Graduate degree	318	43.4		
# children (age ≤16)			0.41	0.81
# workers			0.51	0.71
# seniors (age 65 +)			0.06	0.28
Income: $ 0–50k	64	8.7		
$50–75k	125	17.0		
$100–150k	196	26.6		
$150k +	135	18.3		
Missing	55	7.5		
Multifamily home	148	20.6		
Lived in home: 0–5 years	306	42.6		
# cars			1.74	1.03
# bicycles			2.46	2.03
Car-share member	173	23.8		
Bike-share member	70	9.6		
Transit pass	307	42.2		
# commute days			4.62	0.89
# hours worked			42.34	10.25
Flexible work schedule	451	62.8		
Self-employed	33	4.6		
Traveler perceptions				
Typical travel time: Dissatisfied	239	34.2		
Ideal travel time (min)			13.70	8.76
Teleportation: No	261	37.5		
Travel usefulness: Mostly wasted	81	11.8		
Somewhat wasted	128	18.7		
Somewhat useful	176	25.7		
Mostly useful	157	22.9		
Travel liking: Disliked	106	15.6		
Somewhat liked	238	35.1		
Strongly liked	189	27.8		

commuters and auto drivers/passengers were often grouped together.) If modal interactions were not significant for a given activity, that activity was included in the model with no interaction terms.

Nonlinear representations of travel time were examined but rejected; a quadratic term was not statistically significant, and adding the natural log of travel time only marginally improved the model fit. Interacting travel

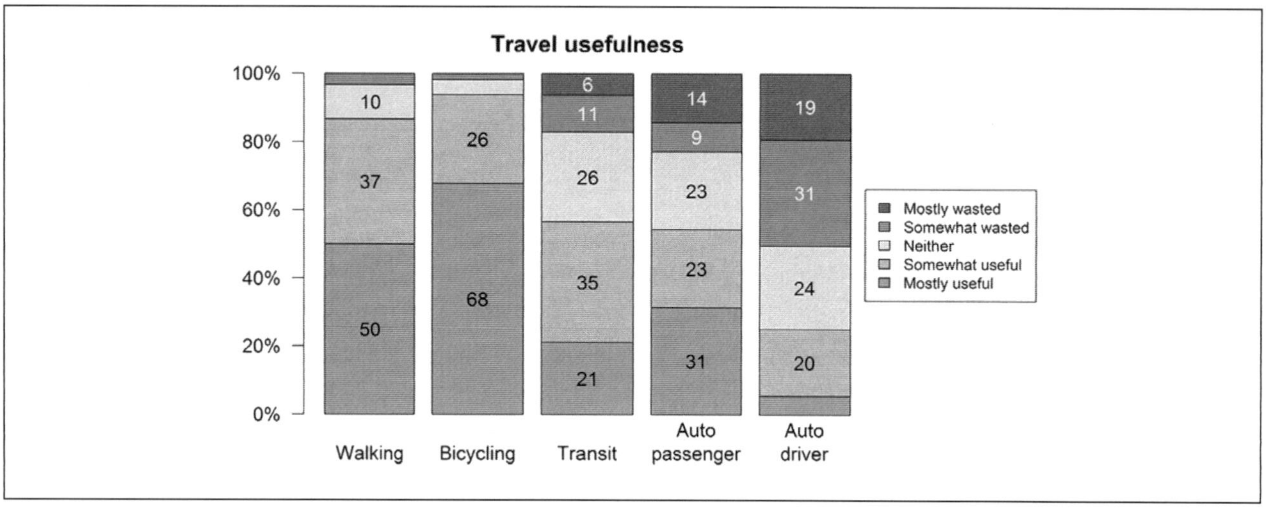

Figure 1. Travel usefulness by commute mode.

time with commute mode in model A yielded a modest but significant reduction in model deviance ($\Delta G^2 = 10.7$, $\Delta df = 4$, $p < 0.05$); the generic travel time coefficient was negative but not statistically significant. Model B did not include the travel time × mode interaction because it was not statistically significant ($\Delta G^2 = 5.6$, $\Delta df = 4$, $p > 0.10$).

All traveler characteristics variables shown in Table 2 and all travel activities and travel activity × mode interactions were included in the models, but only marginally significant ($p \leq 0.10$) coefficients for all variable types (except travel time) are shown. (Complete model results are available from the author.) The full models—including trip characteristics, travel activities, and traveler characteristics—significantly reduced the deviance of their corresponding models containing only category thresholds ($\Delta G^2 = 545.7$, $\Delta df = 68$, $p < 0.001$ for the activity participation model; $\Delta G^2 = 521.0$, $\Delta df = 62$, $p < 0.001$ for the activity duration model). McFadden's pseudo-R^2 goodness-of-fit values were approximately 0.27 for both models.

Some results regarding the association between travel-based multitasking and travel usefulness were consistent, whereas others differed slightly between the two models. Modal interactions with all travel activities were tested, but only a few significant differences were found. In both models, auto travelers who talked with people they knew reported more useful commutes, whereas those reporting "doing nothing" rated their commutes by auto to be less useful. Furthermore, those who sung while driving had more useful commutes. In the activity participation model A, walk and bicycle commuters who were doing passive activities (viewing scenery, watching people, thinking, or daydreaming) had less useful commutes, as did bicycle riders and auto drivers who were planning or navigating the trip. In the activity duration model B, transit commuters who were engaged in passive activities reported more wasteful commutes, as did auto drivers who reported more time spent personal grooming. Finally, among walk and bicycle commuters, exercising for longer durations was assessed to be more useful. Notably, many common activities were not significantly associated with travel usefulness in either model: ICT-related activities, eating or drinking, and listening to music. Reading print newspapers or books while riding public transit had positive and nearly marginally significant coefficients (not shown) in both models.

Differences between the two models were also found for travel time. In model A with mode-specific travel times, longer commutes were more wasteful for auto passengers but more useful for people walking; the travel time coefficient for bicycling was also positive but not significant. In model B, the generic travel time coefficient was negative but not significant. When this interaction was specified (model not shown), the coefficient for auto drivers was negative and marginally significant, and the positive coefficient for walking approached significance. More interestingly, adding this interaction made the coefficient on the exercising activity no longer significant, suggesting high correlations between times spent walking or bicycling and time spent exercising or being physically active.

After controlling for mode-specific activity participation, travel time (by mode), and traveler characteristics, modal differences remained. Bicycling and walking commuters as well as auto and transit passengers all found their commutes to be more useful than did auto drivers; bicycling remained the most useful mode. Commuters traveling on a day with rain reported slightly lower levels of usefulness, although this was not significant in model A.

Table 3. Results of Ordered Logit Models of Travel Usefulness

	Model A with activity participation			Model B with activity duration		
Variable	*B*	*SE*	*p*	*B*	*SE*	*p*
Trip characteristics & weather						
Mode: Walk	1.899	0.929	0.041*	1.779	0.518	0.001*
Bicycle	3.448	0.847	0.000*	2.481	0.418	0.000*
Auto, passenger	2.303	0.810	0.005*	1.423	0.476	0.003*
Transit	1.236	0.534	0.021*	1.898	0.362	0.000*
Travel time (min)	–	–	–	−0.009	0.007	0.225
Travel time × Walk	0.041	0.023	0.070~	–	–	–
Bicycle	0.024	0.016	0.152	–	–	–
Auto, driver	−0.008	0.007	0.275	–	–	–
Auto, passenger	−0.045	0.021	0.031*	–	–	–
Transit	−0.005	0.008	0.503	–	–	–
Day precipitation ≥0.10 in				−0.379	0.198	0.056~
Travel activities (participation or duration)						
Passive activities × Walk & Bicycle	-1.793	0.664	0.007*	–	–	–
Transit	–	–	–	-0.015	0.006	0.019*
Talking with people you know × Auto, driver & Auto, passenger	0.807	0.379	0.033*	0.035	0.018	0.057~
Personal grooming × Auto, driver	–	–	–	-0.071	0.033	0.031*
Singing; dancing × Auto, driver	1.097	0.413	0.008*	0.086	0.024	0.000*
Exercising; being physically active × Walk & Bicycle	–	–	–	0.030	0.012	0.013*
Planning or navigating this trip × Bicycle & Auto, driver	−0.665	0.306	0.030*	–	–	–
Doing nothing × Auto, driver & Auto, passenger	−1.429	0.419	0.001*	−0.079	0.035	0.025*
Traveler sociodemographics						
Age: 18–34 years	−0.531	0.289	0.067~	−0.567	0.295	0.055~
# children (age ≤16)	0.189	0.110	0.087~			
Income: $100–150k				−0.408	0.236	0.084~
# commute days	0.214	0.105	0.042*	0.219	0.106	0.040*
Traveler perceptions						
Typical travel time: Dissatisfied	−0.391	0.202	0.053~	−0.435	0.204	0.034*
Teleportation: No	0.592	0.179	0.001*	0.600	0.179	0.001*
Travel liking: Disliked	−0.603	0.278	0.030*	−0.681	0.284	0.017*
Somewhat liked	0.962	0.219	0.000*	1.037	0.224	0.000*
Strongly liked	2.324	0.272	0.000*	2.298	0.274	0.000*
Model fit statistics						
Sample size (*N*)	642			619		
Deviance (thresholds model), *df*	2,021.7	638		1,944.8	615	
Deviance (full model), *df*	1,476.1	570		1,423.8	553	
Likelihood ratio test (ΔG^2), Δdf, *p*	545.7	68	0.000*	521.0	62	0.000*
McFadden's pseudo-R^2	0.270			0.268		
Trip characteristics only	0.155			0.145		
Travel activities only	0.109			0.101		
Sociodemographics only	0.042			0.042		
Perceptions only	0.162			0.162		

Note: Statistical significance: *= $p \leq 0.05$, ~= $p \leq 0.10$. Coefficients with $p > 0.10$ for all variables (except travel time) not shown. – Independent variable not included in model. Threshold parameters not shown.

Few sociodemographic characteristics were significant predictors of travel usefulness in both models. Age appeared to be positively associated with travel usefulness: younger travelers (aged 18–34) were more likely to report their travel time as being less useful. Those who commuted more frequently viewed their travel time as being more useful.

Instead, traveler perceptions were more strongly associated with travel usefulness. Travelers who were more satisfied with their typical commute travel time and those who would rather not teleport to work reported more useful commutes. Notably, travel liking and travel usefulness on a recent commute trip were strongly and positively related. Given these significant associations, models were also estimated without traveler perceptions variables; results (not shown) were similar with some key exceptions. In both models, mode-specific constants had a larger magnitude, as did travel time coefficients: In model A, travel time was a negative and significant variable for all motorized modes; in model B, the generic travel time coefficient was negative and significant. For travel activities, there were no changes to the significance or signs of modal interaction terms. For traveler sociodemographics, in both models, people living in low-income households, owning fewer bicycles, and those with flexible work schedules were more likely to report making better use of their travel time.

Discussion

Determinants of Travel Usefulness

This research provides insights into subjective assessments of the usefulness of time spent traveling. Consistent with recent studies, about 12% of commuters viewed their travel time as mostly wasted, whereas about twice as many (23%) thought their commutes to be mostly useful. Modal differences in travel usefulness were similar to those found in one previous multimodal study (*17*): most active mode (walk, bicycle) travelers rated their commutes as useful, and more than half of riding mode (transit, auto passenger) users had at least somewhat useful commutes; but around half of auto drivers considered their commutes to be wasted time. These conclusions held even after controlling for travel time, activity participation, and traveler characteristics.

Few traditional traveler attributes were associated with travel usefulness: a model with only sociodemographic characteristics had very low goodness-of-fit ($R^2 = 0.04$). As in previous research, age was a positive factor: younger travelers were more likely to consider their commutes to be wasted time. The lack of significance of most demographic and socioeconomic traveler characteristics foretells difficulties in predicting how useful people will consider their commutes to be without asking them directly.

Instead, traveler attitudes and perceptions seem more closely tied to subjective valuations of travel time, as has been found in the past (*17*). People who were more dissatisfied with their typical commute travel time reported more wasteful commutes, whereas people who did not want to teleport to work had more useful commutes. The inclusion of these variables in the model raises questions about endogeneity: Are people less satisfied with their typical travel times because they do not make productive use of them? Do people prefer not to teleport because they find some aspects, like activity participation, to be useful? A more complex analysis utilizing (for instance) longitudinal structural equation modeling could help to illuminate these potential bidirectional effects but is beyond the scope of this paper and this dataset.

Despite these valid concerns, there are reasons why including such perceptions into a model of travel usefulness may be an appropriate choice. First, a large amount of the variation in travel usefulness (roughly 75% of the thresholds-only model deviance) remained unexplained by the independent variables, suggesting that the association with these perceptions is not strongly deterministic. Second, the perception questions (about travel time satisfaction and the teleportation test) asked specifically about a general condition (a traveler's typical commute), whereas the travel usefulness question asked about a specific case (a traveler's most recent commute trip to work). The consideration of time precedence (*26*) suggests that the general condition might cause the specific case but not the other way around. Of course, with cross-sectional data, the endogeneity issue may empirically remain. In fact, it could be likely that, for a frequently repeated travel behavior like commuting, people have the opportunity to equilibrate their prior perceptions about the usefulness of travel and their satisfaction with travel time such that their perceptions fall more in line with their day-to-day experiences. These issues suggest the need to conduct longitudinal studies examining changes in perceptions of travel usefulness.

The travel liking variable was included in the model as a control variable and to specifically address a concern noted in the literature review: that self-reports of the productive time use benefits of travel-based multitasking may have been confounded with enjoyment of the travel experience, as measured by travel liking. In short, people may have reported a useful commute in part because they liked it. Indeed, the model results provide evidence that this may have been the case: travel liking was positively and significantly related to travel usefulness. Further, the positive effect of singing or dancing—activities that have perhaps the least instrumental or traditionally productive value to travelers—suggests that travelers may have been conflating travel usefulness with travel enjoyment. Nevertheless, this issue may be somewhat exaggerated. Conflation could have happened in the opposite direction too: people may have considered the value of their activity participation when considering how much they liked the trip. This bidirectional effect is perhaps likely, given that the travel usefulness question was asked prior to the travel liking question and prior to all other questions about the travel experience. Improvements to the way in which travel usefulness (and liking) questions are asked could help to further distinguish the two concepts.

The Usefulness of Travel-Based Multitasking

The issues raised in the preceding paragraphs suggest the need for more research on travel usefulness and other subjective assessments of travel-based multitasking. Different question wordings or orders could be tested to examine which one best measures the desired construct (the value, worth, or usefulness of activity participation during travel). The finding that many common activities did not significantly predict travel usefulness may suggest that different kinds of questions are needed, perhaps those asking more about the quality of people's effective time use (*21*). The questionnaire used in the current study specifically eschewed the term "productive" based on the assumption that people may not consider valid examples of travel-based multitasking or travel time use (e.g., listening to music, reading for pleasure) to be traditionally productive activities: that is, associated with work or job responsibilities. This assumption could be revisited.

This discussion raises broader but related questions: To what extent is travel-based multitasking valued as a productive use of travel time vs. a way to make otherwise wasted time more tolerable? How much of a PUT is travel-based multitasking? The many activities (whether measured by participation or duration) that were not associated with travel usefulness suggest that instances of travel-based multitasking may not be considered good or productive uses of time, and that the answers to the questions above depend on the activity. Doing some sort of activity appears to be useful: people who reported and spent more time "doing nothing" considered their time to have been less useful. Nevertheless, the activities positively associated with travel usefulness—talking with other people and singing or dancing—could be ways just to pass the time. Other activities may be more about coping with commuting and less about making productive use of travel time (*16*): The significant negative mode-specific effects of passive activities on travel usefulness suggests that thinking or daydreaming, or looking at scenery and people are not considered useful activities. Additionally, the positive association of commuting frequency with travel usefulness suggests that travelers who commute more often (thus, having a larger commute burden) may have a greater incentive to make use of their travel time.

However, even if commuters are really doing most things just to kill time on the commute, these are still instances of a PUT. Activity participation presumably makes travel utility more positive, even if it does not completely outweigh the disutilities of travel time, cost, or effort. Even reducing a small amount of the commuting burden by viewing scenery or talking with a passenger presumably means that travelers are still better off than if they had been doing nothing. From a policy perspective, although travel-based multitasking may be unlikely to generate travel, it may instead diminish commuters' incentives for reducing travel (*3*, *27*). In addition, perhaps travelers truly do benefit from the transition time or time-out (*5*) provided by conducting activities like thinking or daydreaming, or sleeping or snoozing, but do not fully acknowledge the usefulness of those activities.

One travel activity that appears to be useful—exercising and being physically active—warrants further discussion. As an activity, exercising is unique because it is so highly correlated with active travel modes: most reports of exercising were among people walking and bicycling, and most walk and bicycle commuters reported being physically active for most of the time (*16*). This study provides evidence to suggest that exercising is likely considered a useful form of travel-based multitasking. Notably, people walking and bicycling had the most useful commutes. According to the activity duration model of travel usefulness, people spending more time exercising or being physically active were more likely to have useful commutes. Although exercising was not significant in the activity participation version, the travel time coefficient for walking was positive and there were positive residual associations for both walking and bicycling modes. As noted in the results section, the high covariance between these factors makes representing active travel modes and the exercising activity in a model of travel usefulness more challenging.

An additional (conceptual and empirical) issue with the exercising activity is the difficulty distinguishing the useful benefits of travel-based multitasking from the enjoyable benefits related to travel liking, and distinguishing both types of benefits from the traveling itself (walking, bicycling). In the commuting survey, most people walking and bicycling reported useful commutes but also that they liked their commutes. Singleton and Mokhtarian (*3*) discuss some of the arguments for and against classifying exercising as a travel activity instead of a beneficial part of the travel experience. For instance, exercising on the commute can substitute for nontravel physical activity like going to the gym. On the other hand, walking and bicycling can be a fun activity or a way to improve physical and mental health. More broadly, exercising and doing other activities (e.g., listening to music, viewing scenery, daydreaming) also can help to facilitate the enjoyment of the travel experience. Together, these activities could be contributing to the travel usefulness vs. travel liking conflation discussed earlier. One conceptual solution to the dilemma is to say that the doing (exercising) is the multitasked and useful activity, and that any positive emotions, fulfillment, or liking that may result are part of the travel experience. Nevertheless, if it is difficult for scholars to conceptually distinguish the two, how should we expect survey respondents to make such a distinction? Future research should tackle the empirical challenges of measuring and modeling the subjective usefulness of exercising and physical activity.

Limitations and Future Work

Several limitations of this study are about the data collection effort itself—not studying multiple trip purposes, distinguishing wait and access/egress activities, surveying long-haul transit commuters, or including more detailed questions—and have been discussed elsewhere (*16*). From an empirical perspective, more research is needed to examine whether activity participation or duration (or some other measure of quantity) is more closely linked to the usefulness of travel. Fundamentally, issues with the travel usefulness question suggest a need to continue searching for better measures of this subjective concept of the productivity, usefulness, quality, worth, or value of travel-based multitasking. Notably, questions about the quality and effectiveness of travel time use for different activities (*21*) are promising and would have nicely complemented the travel usefulness question, but they could not be included due to limitations on the survey's length. Utilizing natural or stated preference choice experiments paired with econometric analysis techniques could yield utility-based estimates of a traveler's underlying trade-offs and willingness-to-pay for multitasking potential. Some research is trying to do this with respect to the effects of travel-based multitasking on mode choice behavior (*22*, *28*), yet studies of this type are still in their infancy.

Implications

This study offers several implications for transportation design, planning, and policy, especially considering advances in transportation technologies. It demonstrates that people walking and bicycling may indeed view as useful the time they spend exercising while commuting. This finding suggests that interventions to increase the use of these nonmotorized modes could make people healthier and more productive in their time use. Even shifting commuters from single-occupancy vehicles to public transit or carpooling could enhance their well-being, allowing people to make more use of time spent traveling, even if that time is spent on less traditionally productive activities like technology use or people-watching.

The results of this study also shed light on the value of travel-based multitasking aspects of the PUT concept. People do not appear to consider engaging in most activities (beside exercise) to be a useful way to spend commuting time. Instead, they may be doing things (talking, reading, listening to music, using smartphones, looking out the window, etc.) more to pass the time than to be productive, at least in a region with few long-distance transit commuters. This suggests that attempts to increase public transit ridership by leveraging travelers' proclivities to multitask with on-board or station-area amenities like tray tables, charging stations, and WiFi may be less effective than anticipated. Certainly, transit commuters have reaped the benefits of recent advances in ICT. However, attempts to harness these ongoing technological changes to modify commuting behavior may not necessarily succeed. In summary, a PUT from travel-based multitasking may be more likely to diminish incentives for reducing travel time (i.e., by switching from riding transit to driving) than to actually overcome an increase in travel time through a travel behavior change (i.e., switching from driving to riding transit). From the perspectives of utility and the VTTS, travel-based multitasking may reduce the disutility of travel (reducing the magnitude of VTTS estimates) without necessarily generating an absolute PUT (switching the sign of VTTS estimates) (*3*).

Insights into travel usefulness also have long-term implications, particularly for our understanding of travel behaviors surrounding the adoption of autonomous vehicles (AVs). Many observers have suggested that, by removing the need to drive (at least part of the time), AVs offer the potential to make traveling much more productive and useful. Auto drivers may be transformed into travelers that look and act more like auto passengers or transit riders, engaging in more frequent and varied types of travel-based multitasking. Travel modelers have begun to create simulations (e.g., *29*) on the assumption that AVs will make traveling more productive and reduce values of travel time savings, thus increasing travel demand and potentially vehicle miles traveled, all while reshaping where and how people live and get around. However, the results of the present study suggest that this sort of productivity-based multitasking may not be highly valued by commuters. Instead, travelers might use their travel time more to do social activities, catch up on rest, or simply zone out. Thus, VTTS reductions from travel-based multitasking may be more modest than projected (*30*). Furthermore, the people who find exercising while commuting to be useful may not want to use AVs at all. Overall, this study's findings related to travel-based multitasking and travel usefulness suggest a need to reconsider the potential impacts of widespread vehicular automation.

Acknowledgments

Thanks to Kelly Clifton, Jennifer Dill, Cynthia Mohr, Liming Wang, and others at Portland State University as well as three anonymous reviewers for their valuable feedback. This work was supported in part by a Doctoral Dissertation Fellowship from the National Institute for Transportation and Communities, a program of the Transportation Research and Education Center at Portland State University; and by a graduate fellowship from the Dwight David Eisenhower Transportation Fellowship Program, a program of the Federal

Highway Administration of the U.S. Department of Transportation.

Author Contributions

The author confirms contribution to the paper as follows: study conception and design: PAS; data collection: PAS; analysis and interpretation of results: PAS; draft manuscript preparation: PAS. All authors reviewed the results and approved the final version of the manuscript.

References

1. Salomon, I., and P. L. Mokhtarian. What Happens When Mobility-inclined Market Segments Face Accessibility-enhancing Policies? *Transportation Research Part D: Transport and Environment*, Vol. 3, No. 3, 1998, pp. 129–140. http://doi.org/10.1016/S1361-9209(97)00038-2
2. Mokhtarian, P. L., and I. Salomon. How Derived is the Demand for Travel? Some Conceptual and Measurement Considerations. *Transportation Research Part A: Policy and Practice*, Vol. 35, No. 8, 2001, pp. 695–719. http://doi.org/10.1016/S0965-8564(00)00013-6
3. Singleton, P. A., and P. L. Mokhtarian. Is There a Positive Utility of Travel? A Review of Concepts, Measures, and Evidence. (Available from the authors).
4. Lyons, G., and J. Urry. Travel Time Use in the Information Age. *Transportation Research Part A: Policy and Practice*, Vol. 39, No. 2, 2005, pp. 257–276. http://doi.org/10.1016/j.tra.2004.09.004
5. Jain, J., and G. Lyons. The Gift of Travel Time. *Journal of Transport Geography*, Vol. 16, No. 2, 2008, pp. 81–89. http://doi.org/10.1016/j.jtrangeo.2007.05.001
6. Mackie, P. J., S. Jara-Díaz, and A. S. Fowkes. The Value of Travel Time Savings in Evaluation. *Transportation Research Part E: Logistics and Transportation Review*, Vol. 37, No. 2, 2001, pp. 91–106. http://doi.org/10.1016/S1366-5545(00)00013-2
7. Pawlak, J., J. W. Polak, and A. Sivakumar. Towards a Microeconomic Framework for Modelling the Joint Choice of Activity–travel Behaviour and ICT Use. *Transportation Research Part A: Policy and Practice*, Vol. 76, 2015, pp. 92–112. http://doi.org/10.1016/j.tra.2014.10.013
8. Ettema, D., and L. Verschuren. Multitasking and value of travel time savings. *Transportation Research Record: Journal of the Transportation Research Board*, 2007. 2010: 19–25.
9. Singleton, P. A., and K. J. Clifton. Exploring the Positive Utility of Travel Through Theory and Simulation. Presented at the 14th International Conference on Travel Behavior Research, Windsor, UK, 2015. (Available from the authors).
10. Kenyon, S., and G. Lyons. Introducing Multitasking to the Study of Travel and ICT: Examining its Extent and Assessing its Potential Importance. *Transportation Research Part A: Policy and Practice*, Vol. 41, No. 2, 2007, pp. 161–175. http://doi.org/10.1016/j.tra.2006.02.004.
11. Keseru, I., and C. Macharis. Travel-based Multitasking: Review of the Empirical Evidence. *Transport Reviews*, Vol. 38, No. 2, 2017, pp. 162–183. http://doi.org/10.1080/01441647.2017.1317048.
12. Guo, Z., A. Derian, and J. Zhao. Smart Devices and Travel Time Use by Bus Passengers in Vancouver, Canada. *International Journal of Sustainable Transportation*, Vol. 9, No. 5, 2015, pp. 335–347. http://doi.org/10.1080/15568318.2013.784933.
13. Lyons, G., J. Jain, and I. Weir. Changing Times–A Decade of Empirical Insight into the Experience of Rail Passengers in Great Britain. *Journal of Transport Geography*, Vol. 57, 2016, pp. 94–104. http://doi.org/10.1016/j.jtrangeo.2016.10.003.
14. Berliner, R. M., A. Malokin, G. Circella, and P. L. Mokhtarian. Travel-based Multitasking: Modeling the Propensity to Conduct Activities While Commuting. Presented at 94th Annual Meeting of the Transportation Research Board, Washington, D.C., 2015.
15. Keseru, I., J. Bulckaen, C. Macharis, J. Minnen, I. Glorieux, and T. P. van Tienoven. Is Travel Time Wasted? Evidence from a Time Use Survey in Flanders, Belgium. Presented at the 14th International Conference on Travel Behaviour Research, Old Windsor, UK, 2015.
16. Singleton, P. A. Multimodal Travel-based Multitasking During the Commute: Who Does What? (Available from the author).
17. Circella, G., J. R. Salgado, P. L. Mokhtarian, and M. Diana. *The Impact of Activities While Traveling on the Subjective Valuation of Travel Time.* Northwestern University, Evanston, IL, 2015. http://www.transportation.northwestern.edu/docs/2015/04.23.Mokhtarian.presentation.pdf.
18. Lyons, G., J. Jain, and D. Holley. The Use of Travel Time by Rail Passengers in Great Britain. *Transportation Research Part A: Policy and Practice*, Vol. 41, No. 1, 2007, pp. 107–120. http://doi.org/10.1016/j.tra.2006.05.012.
19. Lyons, G., J. Jain, Y. Susilo, and S. Atkins. Comparing Rail Passengers' Travel Time Use in Great Britain Between 2004 and 2010. *Mobilities*, Vol. 8, No. 4, 2013, pp. 560–579. http://doi.org/10.1080/17450101.2012.743221.
20. Susilo, Y., G. Lyons, J. Jain, and S. Atkins. Rail Passengers' Time Use and Utility Assessment: 2010 Findings from Great Britain with Multivariate Analysis. *Transportation Research Record: Journal of the Transportation Research Board*, 2012. 2323: 99–109.
21. Rosenfield, A., and J. Zhao. Making the Commute Count: Quality of Productive Travel Time Use. Presented at the 56th Annual Conference of the Association of Collegiate Schools of Planning, Portland, OR, 2016.
22. Singleton, P. A. *Exploring the Positive Utility of Travel and Mode Choice.* Doctoral dissertation. Portland State University, Portland, OR, 2017. http://doi.org/10.15760/etd.3447.
23. Venables, W. N., and B. D. Ripley. *Modern Applied Statistics with S*, 4th ed. Springer, New York, NY, 2002.
24. Ory, D. T., and P. L. Mokhtarian. When is Getting There Half the Fun? Modeling the Liking for Travel. *Transportation Research Part A: Policy and Practice*, Vol. 39, No. 2, 2005, pp. 97–123. http://doi.org/10.1016/j.tra.2004.09.006.
25. Russell, M., and P. Mokhtarian. How Real is a Reported Desire to Travel for its Own Sake? Exploring the 'Teleportation' Concept in Travel Behaviour Research.

Transportation, Vol. 42, No. 2, 2015, pp. 333–345. http://doi.org/10.1007/s11116-014-9546-1.
26. Singleton, R. A. Jr., and B. C. Straits. *Approaches to Social Research* (4th ed.). Oxford University Press, New York, NY, 2005.
27. Mokhtarian, P. L. What Good is It? Reflections on the Utility of Travel in a Resource-Constrained Era. Presented at the 8th Annual Martin Wachs Lecture, UCLA Luskin School of Public Affairs, Los Angeles, CA, 2014. https://www.youtube.com/watch?v=RSVWPIyf7C0
28. Malokin, A., G. Circella, and P. L. Mokhtarian. How do Activities Conducted While Commuting Influence Mode Choice? Testing Public Transportation Advantage and Autonomous Vehicle Scenarios. Presented at 94th Annual Meeting of the Transportation Research Board, Washington, D.C., 2015.
29. Childress, S., B. Nichols, B. Charlton, and S. Coe. Using an Activity-based Model to Explore the Potential Impacts of Automated Vehicles. *Transportation Research Record: Journal of the Transportation Research Board*, 2015. 2493: 99–106.
30. Singleton, P. A. Discussing the "Positive Utilities" of Autonomous Vehicles: Will Travelers Really Use Their Time Productively? *Transport Reviews*, 2018. https://doi.org/10.1080/01441647.2018.1470584.

The Standing Committee on Effects of Information and Communication Technologies (ICT) on Travel Choices (ADB20) peer-reviewed this paper (18-00952).

Article

TRR
JOURNAL OF THE TRANSPORTATION RESEARCH BOARD

Transportation Research Record
2018, Vol. 2672(50) 23–34
© National Academy of Sciences:
Transportation Research Board 2018
Article reuse guidelines:
sagepub.com/journals-permissions
DOI: 10.1177/0361198118774671
journals.sagepub.com/home/trr
SAGE

Analyzing Passenger Travel Demand Related to the Transportation Hub inside a City Area using Mobile Phone Data

Gang Zhong[1,2,3,4], Jian Zhang[1,2,3,4], Linchao Li[1,2,3,4], Xiaoxuan Chen[5], Fan Yang[1,2,3,4], and Bin Ran[1,2,3,4]

Abstract

The passenger transportation hub plays a crucial role in the urban transportation system. Analyzing transportation hub related travel demand is necessary to support urban transportation planning and management. However, it is difficult to use the traditional travel survey methods to study travel demand because tracking passenger travel trajectories is a near impossible task. The location information from the cellular system provides a feasible way to solve the problem. This paper concentrates on applying mobile phone data to study passenger travel demand related to the Hongqiao transportation hub in Shanghai, China. First, a method is introduced to collect passenger travel information related to the hub from mobile phone data. Then, travel demand indexes are presented to characterize the travel demand in a visual way. Finally, transportation corridors, which connect the hub and other urban areas, are identified to analyze the distribution of travel demand more thoroughly. The results illustrate that the passenger travel demand shows an obvious tide pattern in the city area with the Hongqiao transportation hub as the center. Moreover, there are two identified transportation corridors which reveal the major distribution directions of the passengers, that is, the city center and the Zizhu industrial development zone. The approach in this study testifies that mobile phone data has great potential for transportation planning and management related to transportation hubs.

Travel demand analysis has long been a vital research topic for transportation planners and administrators. Travel surveys, like the questionnaire and the telephone interview, are usually used to collect the data for a travel demand study. However, the data can't be updated frequently because the processes of these methods are time-consuming and labor-intensive. In recent years, mobile phone data has emerged as a new data source for studying the trip information of urban users conveniently and timely. As involuntary location information is generated by all users, the sample of mobile phone data is huge and unbiased. The advantages have indicated that mobile phone data is an excellent data source for studying macroscopic travel demand including traffic state (*1*), origin–destination (OD) information (*2*), and so forth.

Passenger transportation hub is one of the most important transportation infrastructures inside the urban area, especially when the hub combines several travel modes. Passenger movements related to the hub can appear to affect the traffic state of the city transportation networks. Analyzing passenger travel demand is meaningful to support and evaluate the planning of the hub and the related traffic facilities. Mobile phone data provides a feasible way to extract passenger travel trajectories and analyze the related travel demand. Besides, transportation corridors can be used to display the distribution pattern of the travel demand in a visual way. Smith (*3*) considered a corridor as a "travel shed", that is, an area where trips tend to cluster in a general linear pattern with feeder routes in a metropolitan area. To be more specific, in this study, a corridor is defined as the line linking an OD pair which presents the main

[1]Jiangsu Key Laboratory of Urban ITS, School of Transportation, Southeast University, Jiangsu, China
[2]Joint Research Institute on Internet of Mobility between Southeast University and University of Wisconsin-Madison, Jiangsu, China
[3]Jiangsu Province Collaborative Innovation Center for Technology and Application of Internet of Things, Jiangsu, China
[4]Jiangsu Province Collaborative Innovation Center of Modern Urban Traffic Technologies, Jiangsu, China
[5]TOPS Laboratory, Department of Civil and Environment Engineering, University of Wisconsin-Madison, Madison, WI

Corresponding Author:
Address correspondence to Jian Zhang: jianzhang@seu.edu.cn

distribution pattern of the passenger flow in the city area. The OD information obtained from mobile phone data can be used to identify the transportation corridors connecting the transportation hub and other city areas.

The objectives of this study are to develop a complete methodology to analyze the passenger travel demand related to the transportation hub inside the city area and display the travel demand in a visual way.

Literature Review

Travel Demand Analysis

In recent years, there have been many studies focusing on travel demand analysis using mobile phone data. To study trip generation, algorithms were proposed to identify from mobile phone data the locations where users stopped for specific activities (*4*, *5*). Thus, the travel trajectories of users could be obtained. Based on the collected trip information, parameters including the trip frequency and the trip length were usually extracted to characterize the users' travel behavior and compared to the results of the travel survey (*6*, *7*). To understand trip generation more clearly, home and workplaces were identified for mobile phone users based on certain rule-based algorithms. Thus, the trips of each user could be further classified into Home-Based-Work (HBW) trips, Home-Based-Other (HBO) trips, Non-Home-Based (NHB) trips (*8*, *9*). As for trip distribution, researchers mapped the staying locations identified from mobile phone data into traffic analysis zones to generate an OD matrix in the urban area (*10–12*). Moreover, the travel patterns of urban users were also discovered with the help of cellular information (*13*, *14*). In a recent study (*15*), Dong et al. introduced several normalized indexes of travel generation, attraction and tide pattern to characterize urban travel demand patterns. The method was suitable for dynamic analysis of the travel demand, which is inspiring. However, the previous studies are basically focused on travel demand analysis in the whole urban area, barely considering some important infrastructures. This study attempts to analyze the travel demand related to transportation hubs in the urban area which are crucial for urban transportation planning.

Transportation Corridor Identification

OD matrix information was usually used as the basic data to identify the transportation corridors. Three major identification methods were proposed by previous studies including graph theory (*16*), cluster analysis (*17*), and projection model (*18*, *19*). Considering the complexity of the algorithms and the integrity of the parameters, the projection model was the better identification method. Chen et al. proposed the OD projection model by calculating the attraction volumes between OD pairs and identified transportation corridors by analyzing the similarity between OD pairs using a clustering method (*18*). Wang et al. further tested the method with the OD matrix data collected from mobile phone data and proved it is applicable in the urban area (*19*). However, the method is not suitable for a much smaller area inside the city (such as the hub area), because the distributions of the OD pairs related to the area are concentrated with the area as the center. Although the method can identify the superior corridors (which represent the major distribution patterns of the passenger flow), it is easy to neglect the inferior corridors (which represent the minor distribution patterns of the passenger flow) because of the strict cluster criteria. The inferior corridors are also important for the transportation planning and management related to the area. The method has been revised in a recent study (*20*) to avoid this disadvantage. In this paper, the revised method is used to help understand the passenger travel demand more comprehensively.

Data Description

In this study, the Hongqiao transportation hub is chosen as the study area which is in the city of Shanghai, China. It is an international passenger distribution center consisting of multiple traffic modes such as aviation, railway, metro, bus, and so forth. The average daily passenger volume was around 684,300 in 2013 according to the official data (*21*). The hub is located outside the central urban area, as shown in Figure 1. The main city center is also marked on the figure. The traffic analysis zone (TAZ) information in Shanghai was also collected as the basic data for the travel demand analysis, shown as the blue grids in Figure 1.

The mobile phone data was collected from one of the major communication operators in China, covering the entire city area of Shanghai. Each record contains several segments including a user ID, a location area code (LAC), a cell ID, a timestamp, and an event ID. The user ID is encrypted in the dataset because of the privacy policies. Besides calling and texting, there are also some other types of triggering events such as power on/off, handover, location update, and so forth. The two segments, LAC and cell ID, can be used to locate the users at the cellular tower level. There are over 60,000 cellular stations found from the dataset in the city area of Shanghai. To understand the spatial resolution of the mobile phone data, the distances are analyzed between the cellular towers and their closest neighbor cellular towers, that is, the spacing gaps. The results explain that the spacing gaps range from under 50 m in the downtown area to over 1,500 m in the suburb districts. In general, over 88% of the spacing gaps are less than 200 m and only around 1.7% of the spacing gaps are larger than 500 m.

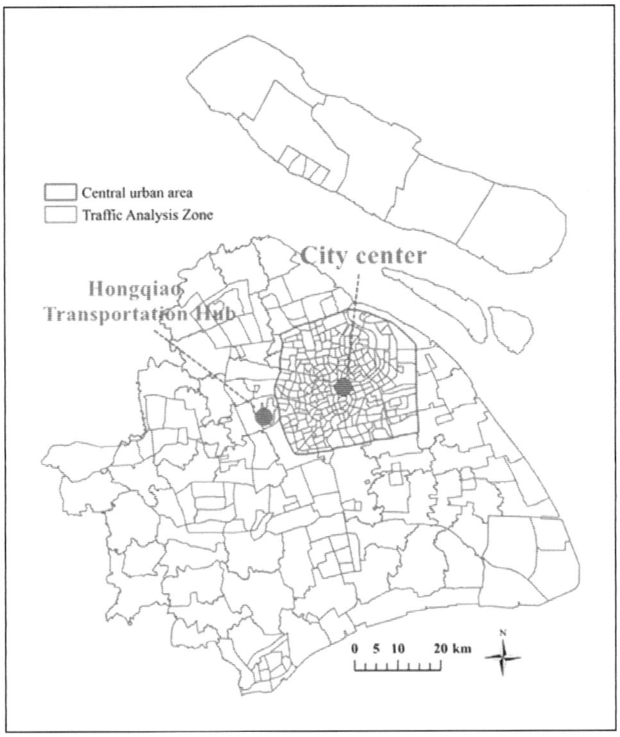

Figure 1. Geographic information of the study area and the TAZs.

In a recent study (*22*), a method was proposed to identify hub passengers from mobile phone users and classify the passengers into several types. Based on the previous study, the mobile phone data of the identified passengers on November 20, 2013 (a workday) and November 24, 2013 (a weekend) are used in this research to analyze the travel demand related to the transportation hub inside the urban area. The mobile phone data has been preprocessed to erase potential noise data such as duplicated data, ping-pong data, drift data, and so forth.

Methodology

Extracting Travel Information of the Hub Passengers

Trips Generation. To collect the travel information, it is necessary to transfer the mobile phone data into trip data used for travel research. In previous studies (*4*, *5*), spatial and temporal thresholds were used to generate the locations where users stop for specific activities. However, the spatial threshold was usually applied to restrict the distances between the mobile phone points (i.e. the coordinate locations of the mobile phone records). Chen et al. (*23*) presented an interesting conclusion in their study indicating that staying points in one place were more likely to be distributed in all directions, whereas the moving points were more likely to be distributed along a linear line. Inspired by their conclusion, a revised method is proposed to extract the staying locations from the mobile phone data. The points in one staying location can be enclosed by a limited radius circle considering the distribution characteristics of the mobile phone points. Therefore, the spatial threshold can be replaced as the restriction for the smallest radius of the qualified enclosing circles.

$P_a = \{p_1^a, p_2^a, \cdots, p_k^a, \cdots\}$ is used to denote the data records set of user a, where p_k^a is the kth location point $p_k^a = (x_k^a, y_k^a, t_k^a)$, (x_k^a, y_k^a) is the coordinate information, and t_k^a is the timestamp. For a segment of the observation series when user a stays to engage in an activity $\{p_m^a, p_{m+1}^a, \cdots, p_n^a\}$, the filtering method is denoted as

$$\text{min\_R}(p_m^a, p_{m+1}^a, \cdots, p_n^a) < R\_th \quad (1)$$

$$t_n^a - t_m^a > t\_th \quad (2)$$

Where $\text{min\_R}(p_m^a, p_{m+1}^a, \cdots, p_n^a)$ is the minimum radius of the enclosing circles for the staying points, R_th is the spatial threshold for the radius of the enclosing circles, t_th is the temporal threshold.

The threshold values are important in the trips generation method. It is necessary to use sensitivity analysis to set the threshold values. Three temporal threshold values (10 min, 20 min, and 30 min) and four spatial threshold values (150 m, 250 m, 350 m, and 450 m) are chosen to do this work. The travel trajectories of 400,000 users appearing in Shanghai for 12 continuous days are extracted using the combination of each two threshold values. On November 20, 2013 (workday), the average trip frequency of the users is 2.24 using the group of threshold values (30 min, 350 m) which is closest to the average trip frequency of the residents (2.23) according to the travel survey of residents in Shanghai (*24*). Therefore, in this study, the spatial threshold is set as 350 m and the temporal threshold is 30 min.

The cluster of staying points $\{p_m^a, p_{m+1}^a, \cdots, p_n^a\}$ are assumed to be located at the center of the corresponding minimum enclosing circle as the activity place, $C_\lambda^a = (x_\lambda^a, y_\lambda^a)$. The travel between two consecutive activity places composes a trip of user a, while the trajectory contains the trips in chronological order.

In the study, the first point and the last point of user a on a typical day are set as staying to avoid neglecting some trips of the user, such as the arrival/departure trips in the "Trips filtering" section.

Trips Filtering. To study the travel demand related to the transportation hub inside the urban area, it is necessary to filter the trips of the passengers. The filtering method contains two major parts: the first part is to guarantee that the trips are all inside the city area, while the second part is to select the trips related to the transportation hub.

(1) The first part of trips filtering

In a previous study (*22*), the transportation passengers were divided into two main categories as incity passengers or external passengers. The definitions are listed as follows:

Incity passengers: people who travel inside Shanghai.
External passengers: people who travel between Shanghai and other cities.

The external passengers need external trips to travel between cities, that is, the origins or the destinations of these trips are actually located in other cities. Nevertheless, the generated trips in the "Trips generation" section may contain external trips because the first and the last points of a user are set as staying on a typical day in the trips generation method. For example, if a passenger arrives at Shanghai by railway in the morning, the first trip may be identified as the trip from the city boundary to the transportation hub although it should be an external trip in fact. The situation is similar for the last trip of a passenger departing from Shanghai by ground travel mode in the evening. Therefore, these external trips must be filtered out to guarantee that the trips are all inside the city area.

Considering passengers may not stop in the transportation hub for a long enough time, for an arrival trip, it may end at the transportation hub or pass through the hub and then end at a place inside the city. Hence, this trip should either be deleted or the origin place moved to the transportation hub, as shown in Figure 2*a*. The process for the departure trip is similar and displayed in Figure 2*b*.

(2) The second part of trips filtering

The passengers travel through the transportation hub at least once on the research days, whereas not all their trips are related to the hub. To collect the related trips, the parameter Trip Relevancy (TR) is proposed to quantify the correlation between each trip and the hub. If the trips pass by the hub, or the origins or destinations are in the hub, the trips are directly related to the hub, TR = 1. Besides these trips, passengers may also have some indirectly related trips. For example, a passenger may drop by the office to get some files before taking a business trip through the hub to other cities. The trip to the office is indirectly related to the hub, followed by the directly related trip from the office to the hub. The indirectly related trip is also important to comprehensively analyze the influence range of travel demand related to the transportation hub. For the *i*-th directly related trip of a passenger, the TR of the indirectly related trips before or after this trip can be calculated as

$$\mathrm{TR}_b = \begin{cases} \left(1 - \frac{t_i^f - t_b^d}{T}\right)^2, t_i^f - t_b^d < T \\ 0, t_i^f - t_b^d \geq T \end{cases} \tag{3}$$

$$\mathrm{TR}_a = \begin{cases} \left(1 - \frac{t_a^a - t_i^l}{T}\right)^2, t_a^a - t_i^l < T \\ 0, t_a^a - t_i^l \geq T \end{cases} \tag{4}$$

where TR_b and TR_a are respectively the TR of the trips before and after the *i*-th directly related trip, t_i^f and t_i^l are respectively the timestamps of the first and last mobile phone data record in the transportation hub of the *i*-th directly related trip, t_b^d and t_a^a are respectively the departure and arrival time of the studied indirectly related trip, T is maximum affected time interval for the transportation hub. According to the travel survey in a previous research study (*25*), the maximum time is 175 min for a passenger traveling to the Hongqiao hub inside Shanghai. Therefore, T is set as 175 min in this paper.

The trips with the criterion that TR > 0 are selected to guarantee that the collected trips are all related to the transportation hub. The OD matrix of the collected trips is also obtained using the TR.

$$\mathrm{OD}_{pq} = \sum_j \mathrm{TR}_{pq} + \sum_j \mathrm{TR}_{qp} \tag{5}$$

Where OD_{pq} is the OD volume between TAZ p and TAZ q, TR_{pq} is the total TR of trips from TAZ p to TAZ q of the *j*th user, TR_{qp} is the total TR of trips from TAZ q to TAZ p of the *j*th user.

In order to extend the number of users to the actual number of passengers, the OD volumes are divided by the market share of the communication operator providing the mobile phone data.

General Analysis of the Travel Demand Patterns

In a related study (*15*), Dong et al. defined several indexes to describe the urban travel demand patterns in a visualized and normalized way. Two of the indexes are introduced in the paper which are the traffic generation index (TGI) and the traffic attraction index (TAI).

TGI is used to quantify the capacity of traffic flow from the target TAZs to other TAZs, while TAI characterizes the capacity of traffic flow from other TAZs to the target TAZs. The expressions of the indexes in the original study are listed as

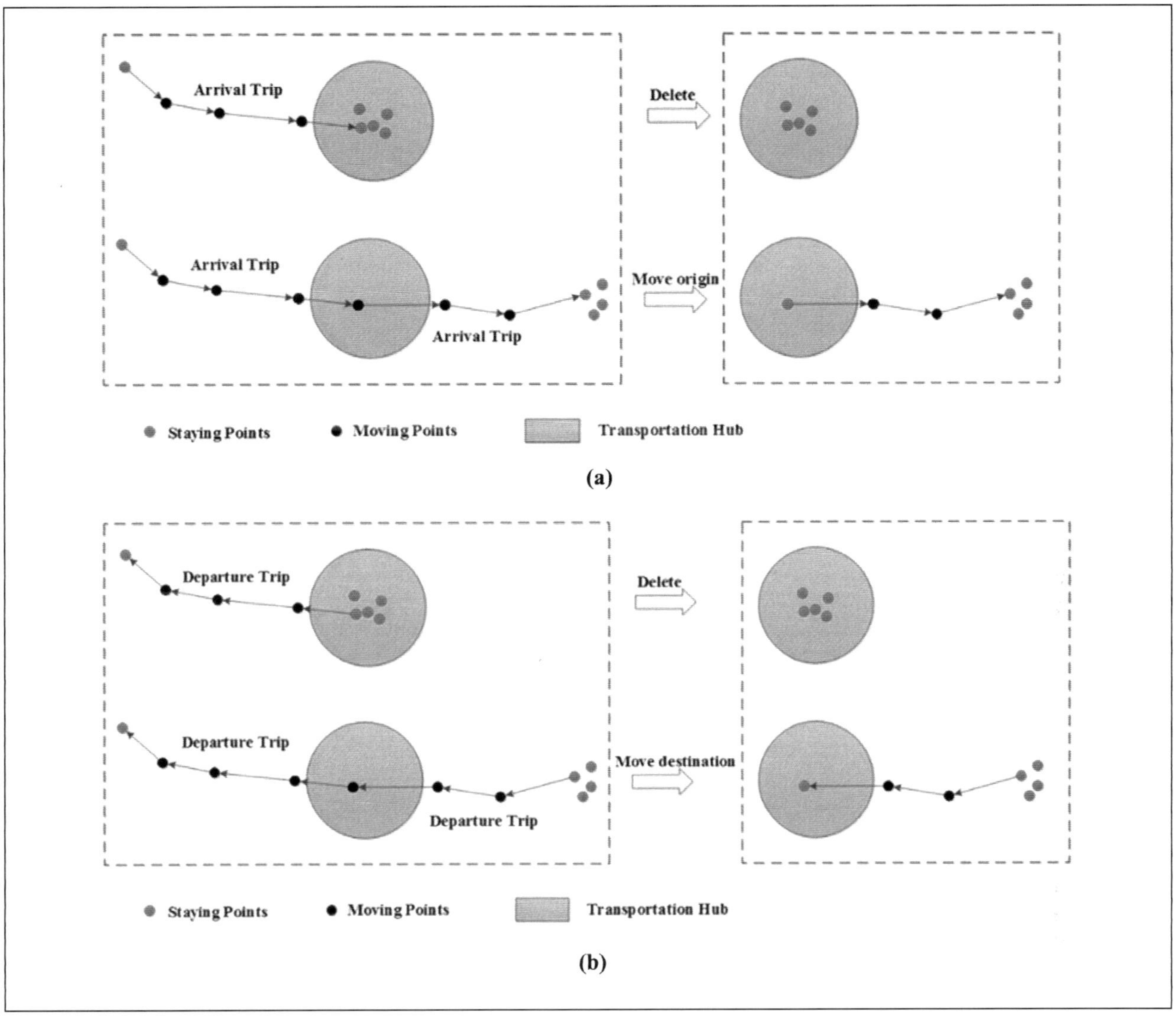

Figure 2. Filtering method for the (*a*) arrival trip and (*b*) departure trip.

$$\mathrm{I\_g}_p^\tau = \frac{V_{\mathrm{out},p}^\tau}{\max\limits_{\tau,k}(V_{\mathrm{in},k}^\tau, V_{\mathrm{out},k}^\tau)} \tag{6}$$

$$\mathrm{I\_a}_p^\tau = \frac{V_{\mathrm{in},p}^\tau}{\max\limits_{\tau,k}(V_{\mathrm{in},k}^\tau, V_{\mathrm{out},k}^\tau)} \tag{7}$$

Where $\mathrm{I\_g}_p^\tau$ and $\mathrm{I\_a}_p^\tau$ are respectively TGI and TAI of TAZ p in the periodτ, $V_{\mathrm{out},p}^\tau$ and $V_{\mathrm{in},p}^\tau$ are respectively the generation and attraction traffic volumes of TAZ p in the period τ.

However, it is biased to directly use generation and attraction traffic volumes in the expressions because the geometric areas of the TAZs are not identical. With the same traffic volume, the two TAZs with quite different geometric areas still have varying travel demands. Therefore, some modifications are needed to make the indexes more appropriate for the study. Instead of the traffic volume, the ratio of the traffic volume to the geometric area is used.

$$\mathrm{IR\_g}_p^\tau = \frac{VR_{\mathrm{out},p}^\tau}{\max\limits_{\tau,k}(VR_{\mathrm{in},k}^\tau, VR_{\mathrm{out},k}^\tau)} \tag{8}$$

$$\mathrm{IR\_a}_p^\tau = \frac{VR_{\mathrm{in},p}^\tau}{\max\limits_{\tau,k}(VR_{\mathrm{in},k}^\tau, VR_{\mathrm{out},k}^\tau)} \tag{9}$$

Where $\mathrm{IR\_g}_p^\tau$ and $\mathrm{IR\_a}_p^\tau$ are respectively the modified TGI and TAI of TAZ p in the period τ, $VR_{\mathrm{out},p}^\tau = V_{\mathrm{out},p}^\tau/A_p$, $VR_{\mathrm{in},p}^\tau = V_{\mathrm{in},p}^\tau/A_p$, A_p is the geometric area of TAZ p, $0 \le \mathrm{I\_g}_p^\tau \le 1, 0 \le \mathrm{I\_a}_p^\tau \le 1$.

TGI and TAI represent the possibilities of trip generation and attraction in the target TAZ respectively. The values close to 1 stand for high possibilities, conversely, the values close to 0 stand for low possibilities.

Identifying the Transportation Corridors

As illustrated at the start of the paper, a corridor is defined as the line linking an OD pair which presents the main distribution pattern of the passenger flow in the city area. It can be used to analyze the spatial connections between the transportation hub and the regions in the city and understand the main trends of the passenger distribution. In a recent study (*20*), a modified method has been proposed to identify the transportation corridors based on the work of Chen et al. (*18*). The brief sketch of the modified method is described as follows. OD link is used to represent the line connecting the center points of the origin TAZ and the destination TAZ in a OD pair.

The OD Projection Model

Step 1: Calculating the attraction volume of each OD link from the surrounding OD links.
Step 2: Defining the correlation index between OD links.
Step 3: Building the correlation matrix with the index between each two OD links.

The OD Cluster Model

Step 1: Generating the cut matrix of the correlation matrix with a given constant.
Step 2: Calculating the membership of each row to a predefined cluster center and clustering the OD links based on the values of the membership.
Step 3: Repeating last step until reaching the stop condition and identifying the predefined center OD link of each cluster as the corridor.

Analysis Results

General Travel Information

The generation and attraction volume of each TAZ are visualized in Figure 3 based on the collected OD information. In the same TAZ, the generation volume and the attraction volume are relatively balanced, while the total volumes in the different TAZs are quite distinct. The TAZs close to the Hongqiao hub tend to have much higher total volumes, especially the adjacent ones in the suburb area, which indicates that the passenger trips may distribute radially with the hub as the center. Moreover, the generation and attraction volume both tend to spread towards the city center. The volumes are quite small in the TAZs near to the north and the south city edges. It is interesting that some TAZs far away from the hub in the east suburb area also have considerable volumes. It is further found that there is an international airport (the Pudong Airport) located in TAZ 400 (shown in Figure 3*a*). The finding manifests the passenger connection existing between the Hongqiao hub and the Pudong airport.

Travel Demand Indexes Analysis

Three time-periods are chosen to analyze the travel demand in the morning peak hours (06:00–10:00), the flat peak hours (10:00–14:00), and the evening peak hours (16:00–20:00).

The results of TGI and TAI on the workday (November 20) are shown in Figure 4. The first row (Figure 4*a–c*) displays the results of TGI through the day, while the second row (Figure 4*d–f*) represents the results of TAI across different time periods. From the figure, there are three conclusions that can be drawn. Firstly, the TAZs covering the Hongqiao hub have quite high TGI and TAI values. Secondly, the importance of the central city area in passenger trips is highlighted in the figure, which eliminates the bias in Figure 3 by using TGI and TAI instead of the generation and attraction volumes. Finally, the passenger connection between the Hongqiao hub and the Pudong airport can be confirmed, because TAZ 400 always has higher results than the surrounding TAZs in the east suburb area of the city.

Besides the conclusions, the change of the results should also be analyzed. Through the horizontal comparison, it can be found that the distribution of TGI becomes more concentrated through the day while the distribution of TAI shows the opposite trend. The vertical comparison illustrates that the distributions of TGI and TAI are relatively balanced in the noon and more unbalanced in the morning and evening. Both comparisons show that the travel demand of the passengers shows a tide pattern. The passengers come from various regions of the city in the morning to take the travel modes in the hub and go towards various TAZs in the evening.

To explore the reason for the finding in Figure 4, the travel demand indexes should be further analyzed for different categories of passengers. In the methodology section, the passengers have been divided into two main categories (the incity passengers and the external passengers). In fact, the external passengers also contain several more specific categories of which the arrival passengers and the departure passengers are the most important ones. The number of the two categories account for nearly 90% of the total number of the external passengers, according to a previous study (*22*). The definitions are as follows:

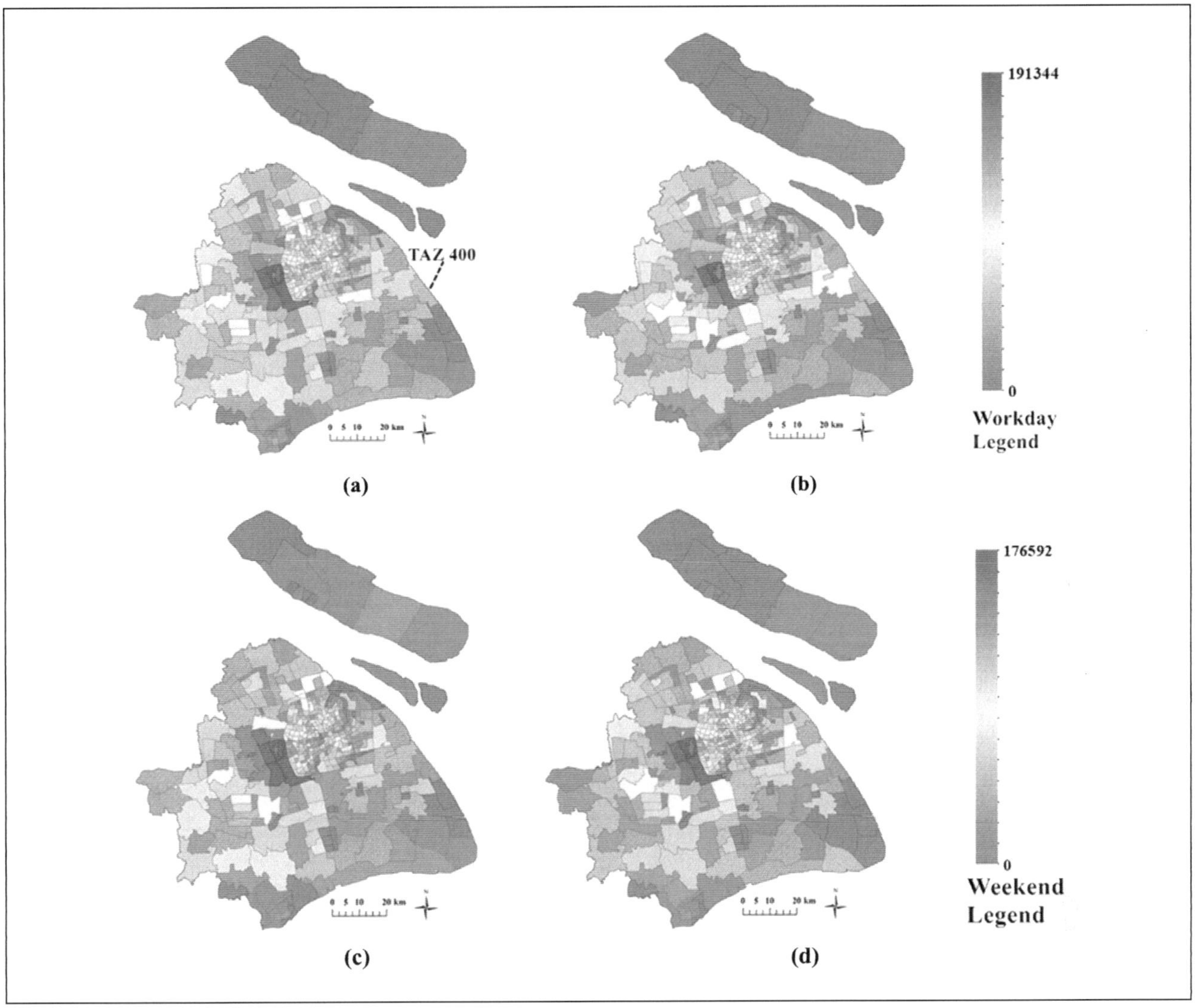

Figure 3. The generation and attraction volumes on each study day: (*a*) generation on the workday, (*b*) attraction on the workday, (*c*) generation on the weekend, and (*d*) attraction on the weekend.

Arrival passengers: the hub passengers who come from other cities and don't leave on the day from the hub.

Departure passengers: the hub passengers who travel to other cities and don't return on the day from the hub.

The results of the travel demand indexes are displayed in Figure 5 for the incity passengers, the arrival passengers, and the departure passengers which are respectively represented as C1, C2, and C3.

From the first row in Figure 5, it can be seen that the travel demand of the incity passengers presents the tide pattern. In the morning, the distribution of TGI spreads in a larger scope while the distribution of TAI tends to be more concentrated in the TAZs between the hub and the city center. In the evening, the situation is the contrary. The tide pattern is quite similar to that of the total passengers.

The second row illustrates that the distribution of C2 TGI is concentrated in the TAZs covering the hub because the arrival passengers disperse from the hub to different places in the city after the arrival. Since the number of the arrival passengers just starts to rise in the morning (according to the results of a previous study (*22*)), the values of TAI in the morning is generally small. The number of the arrival passengers stays at a higher value in the evening, so the range is larger and the values are higher for the distribution of TGI in the evening.

The results of the departure passengers are displayed in the third row of Figure 5. Contrary to the results of

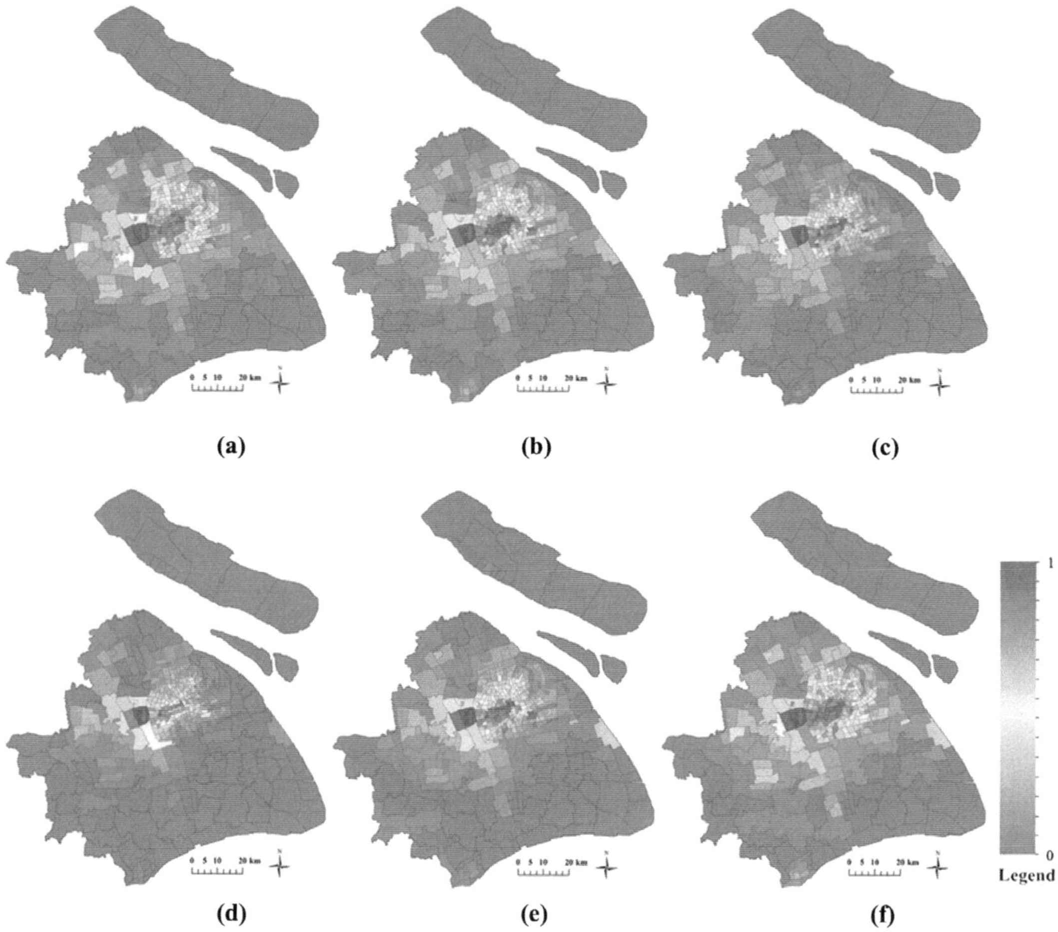

Figure 4. The results of passenger travel demand indexes in the city: (*a*) morning TGI, (*b*) noon TGI, (*c*) evening TGI, (*d*) morning TAI, (e) noon TAI, and (*f*) evening TAI.

the arrival passengers, the trips of the departure passengers tend to start from different areas in the city and end at the area around the hub. Therefore, the distribution of C3 TAI is concentrated in the TAZs covering the hub. The distribution range of TGI in the morning is larger than that in the evening, which is also related to the number of the departure passengers.

Since the findings on the weekend (November 24) are similar, there is no more detailed description of the results on the weekend in this paper.

Transportation Corridors Related to the Hongqiao Transportation Hub

In this part, the OD pairs with volumes over 300 are chosen to identify the transportation corridors. These chosen OD pairs contribute over 73% of the total OD volume and around 0.3% of the total number of the OD pairs. The geographic distributions of the qualified OD pairs on each day are displayed in Figure 6. It can be found from the figure that the OD links distribute radially with the hub as the center which conforms to the earlier hypothesis.

On both research days, there are two identified transportation corridors related to the hub, as shown in Figure 7. The superior corridor and the inferior corridor are defined in the literature review section. Although the results are not exactly the same on the two days for the superior corridor, the general directions are similar which connects the hub and the main city center. When Chen's method (*18*) is used based on the data, only the superior corridor can be identified. However, the revised method (*20*) can identify an extra inferior corridor. The results of the inferior corridor are also similar on the two

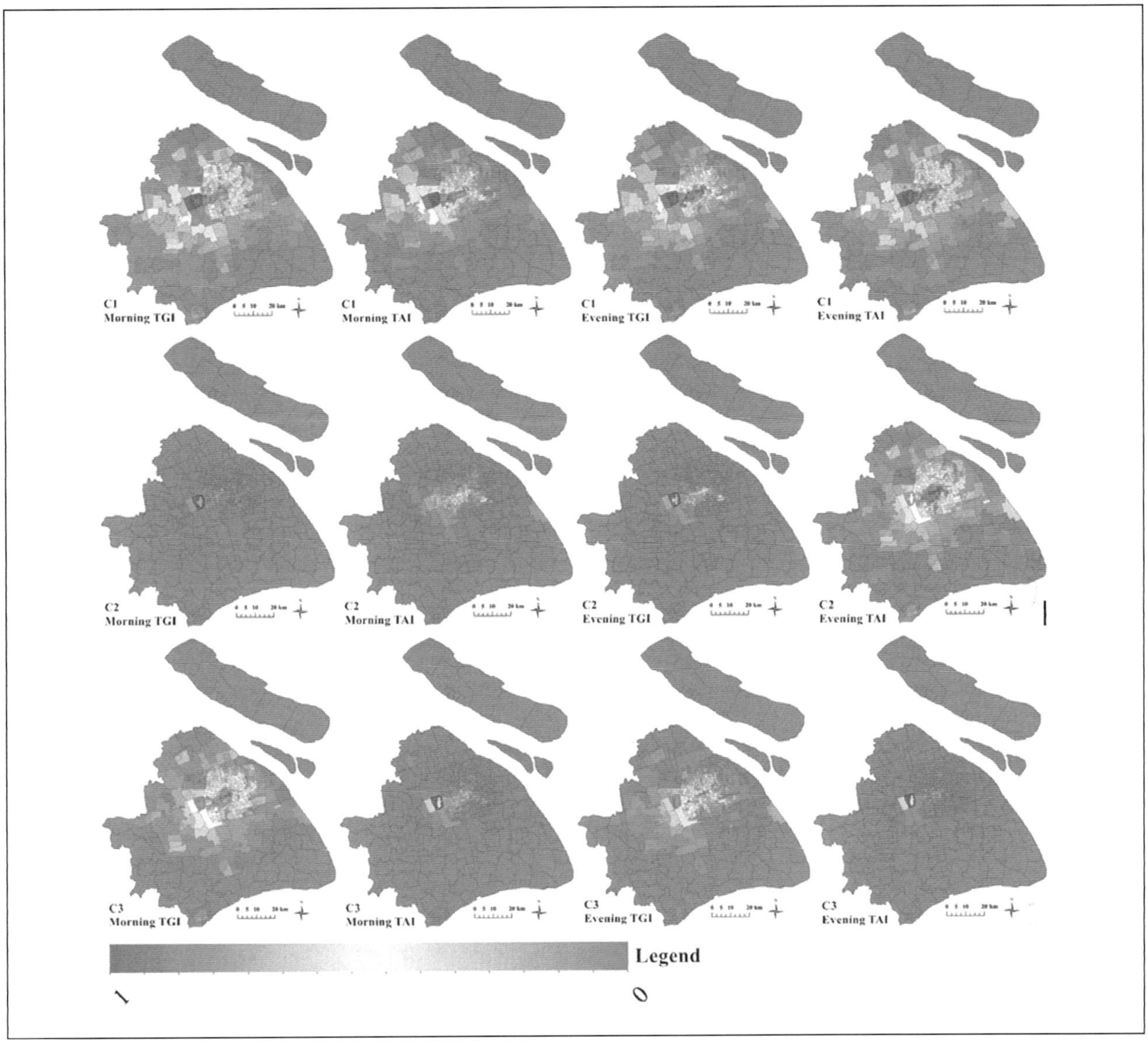

Figure 5. The results of the travel demand indexes for different categories of passengers.

research days, which connects the hub and an important high-tech development zone in Shanghai, the Zizhu industrial development zone.

The transportation corridors reflect the distribution and concentration of the passenger travel demand more specifically. When the corridors are matched to the real transportation networks, several major travel lines can be found for the passengers in the city area, as shown in Figure 7*c*.

The superior corridor: Metro line #2, Metro line #10, and Yan'an elevated road.
The inferior corridor: Jiamin elevated road, Outer ring road, and S4 highway.

Conclusions

This paper presents a complete approach to the study of the hub passenger travel demand in a city area based on mobile phone data. The Hongqiao transportation hub is taken as a detailed application to conduct the approach. There are three main contributions addressed as follows:

(1) The approach provides a feasible way to study the passenger travel demand related to the transportation hub. Travel surveys in the hub perhaps can collect a limited number of passenger trips before arriving at the hub. However, it is almost impossible to trace the passenger movements

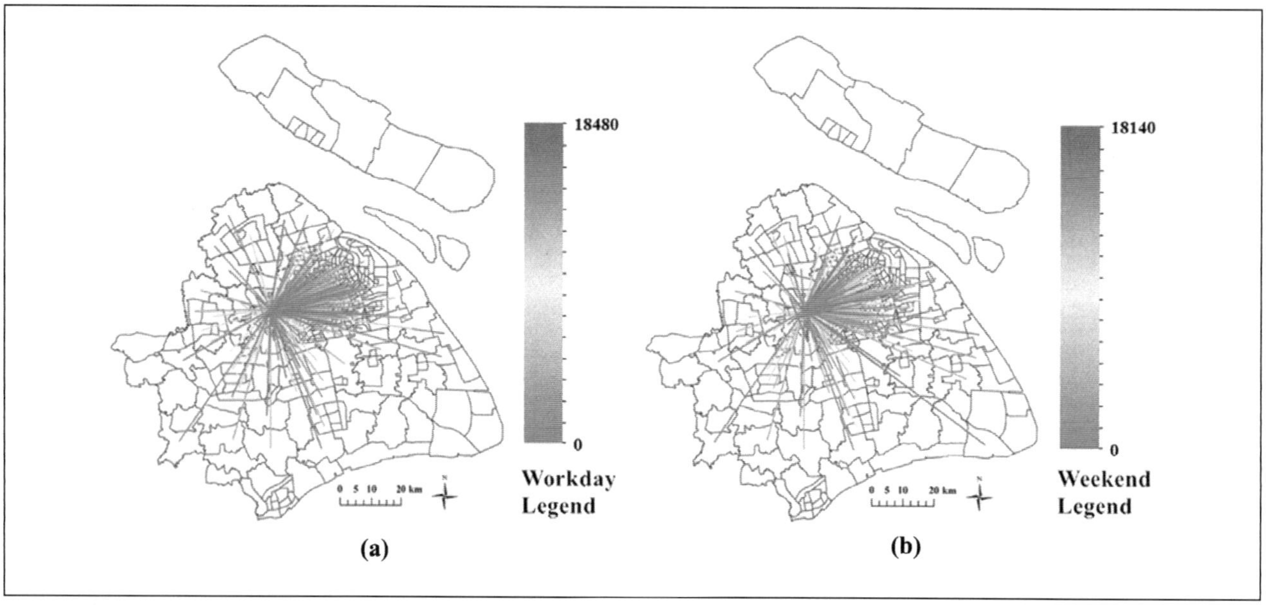

Figure 6. The geographic distributions of the qualified OD pairs on: (*a*) the workday and (*b*) the weekend.

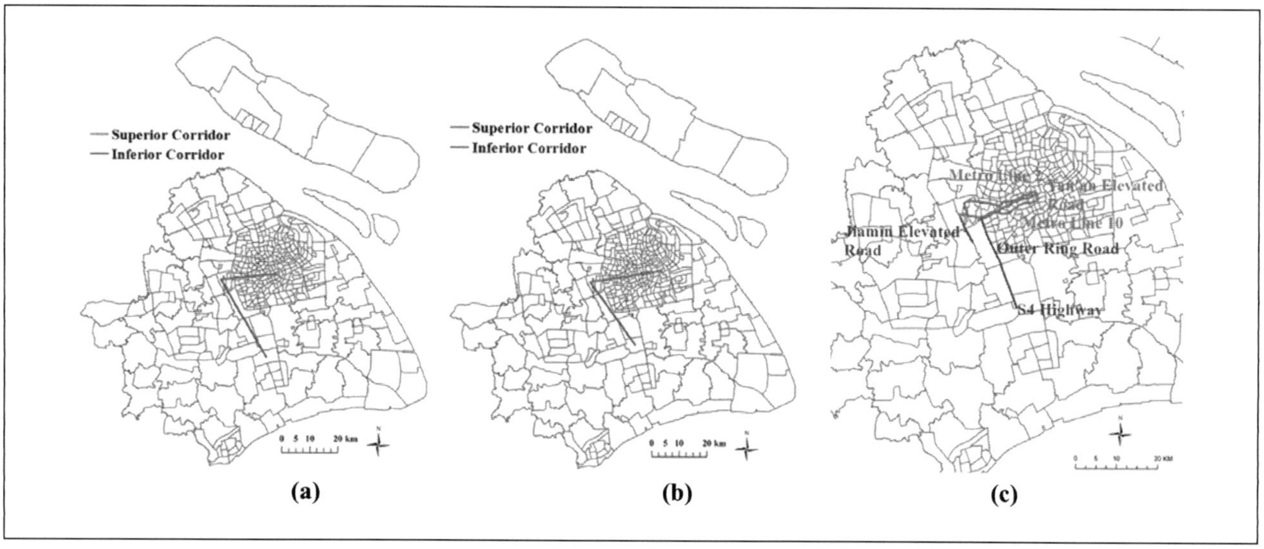

Figure 7. The identified transportation corridors related to the Hongqiao transportation hub: (*a*) results on the workday, (*b*) results on the weekend, and (*c*) transportation network.

after leaving the hub using the traditional method. This paper overcomes this problem by using mobile phone data.

(2) The travel demand analysis can support transportation managers to optimize transportation policy. For example, this paper finds that the passenger travel demand presents an obvious tide pattern related to the Hongqiao hub. Therefore, reversible lanes can be considered as a method to alleviate the pressure on the traffic networks around the hub, especially on the roads matching the transportation corridors.

(3) The approach can also be used to evaluate the operation of the transportation hub by analyzing the passenger attraction range in the city area. The information is helpful to design the development of the hub and plan the related transportation facilities to fulfill the passenger travel demand. For instance, transportation managers should pay attention to the passenger connection

between Hongqiao hub and Pudong airport. Transportation modes connecting the two places are necessary to facilitate the passengers directly.

There are also some vital findings of the travel demand related to the Hongqiao hub.

(1) The passenger travel demand is concentrated on several OD pairs distributing radially with the hub as the center.
(2) The passenger travel demand shows obvious tide pattern. In the morning, the passengers tend to come from various places in the city and travel towards the city center. However the situation is opposite in the evening.
(3) There are two transportation corridors related to the Hongqiao hub. The superior one connects the hub and the city center, while the inferior one connects the hub and the Zizhu industrial development zone.

Future work will continue to explore the approach forecasting the dynamic travel demand related to the transportation hub. It is also an interesting topic to study the travel patterns of different categories of passengers and predict their travel trajectories. To compare with the hub passenger travel demand, the travel demand patterns of users in the entire city area can also be further studied.

Acknowledgments

This study is partially supported by the National Natural Science Foundation of China (No.61620106002, No.71701044), the Information Technology Research Project of Ministry of Transport of China (No.2015364X16030), the Fundamental Research Funds for the Central Universities (No. 2242018K30015), and the Natural Science Foundation of Jiangsu Province (No. BK20150616). The support provided by China Scholarship Council (CSC) during a visit of G. Zhong to UW-Madison is acknowledged.

Author Contributions

The authors confirm contribution to the paper as follows: study conception and design: Gang Zhong, Jian Zhang; data collection: Bin Ran, Jian Zhang; analysis and interpretation of results: Gang Zhong, Xiaoxuan Chen, Fan Yang; draft manuscript preparation: Gang Zhong. All authors reviewed the results and approved the final version of the manuscript.

References

1. Zhang, J., S. He, W. Wang, and F. Zhan. Accuracy Analysis of Freeway Traffic Speed Estimation Based on the Integration of Cellular Probe System and Loop Detectors. *Journal of Intelligent Transportation Systems*, Vol. 19, No. 4, 2015, pp. 411–426.
2. Pan, C., J. Lu, S. Di, and B. Ran. Cellular-Based Data-Extracting Method for Trip Distribution. *Transportation Research Record: Journal of the Transportation Research Board*, 2006. 1945: 33–39.
3. Smith, S. A. *NCHRP Report 435: Guidebook for Transportation Corridor Studies: A Process for Effective Decision-Making*. Transportation Research Board, Washington, D.C., 1999.
4. Asakura, Y., and E. Hato. Tracking Survey for Individual Travel Behaviour Using Mobile Communication Instruments. *Transportation Research Part C: Emerging Technologies*, Vol. 12, No. 3, 2004, pp. 273–291.
5. Phithakkitnukoon, S., T. Horanont, G. Di Lorenzo, R. Shibasaki, and C. Ratti. Activity-Aware Map: Identifying Human Daily Activity Pattern Using Mobile Phone Data. *Proc., International Workshop on Human Behavior Understanding*, 2010, pp. 14–25. https://dx.doi.org/10.1007/978-3-642-14715-9_3.
6. Bekhor, S., Y. Cohen, and C. Solomon. Evaluating Long-Distance Travel Patterns in Israel by Tracking Cellular Phone Positions. *Journal of Advanced Transportation*, Vol. 47, No. 4, 2013, pp. 435–446.
7. Calabrese, F., M. Diao, G. Di Lorenzo, J. Ferreira, and C. Ratti. Understanding Individual Mobility Patterns from Urban Sensing Data: A Mobile Phone Trace Example. *Transportation Research Part C: Emerging Technologies*, Vol. 26, 2013, pp. 301–313.
8. Çolak, S., L. P. Alexander, B. G. Alvim, S. R. Mehndiratta, and M. C. González. Analyzing Cell Phone Location Data for Urban Travel: Current Methods, Limitations, and Opportunities. *Transportation Research Record: Journal of the Transportation Research Board*, 2015. 2526: 126–135.
9. Huntsinger, L. F., and K. Ward. Using Mobile Phone Location Data to Develop External Trip Models. *Transportation Research Record: Journal of the Transportation Research Board*, 2015. 2499: 25–32.
10. Calabrese, F., G. Di Lorenzo, L. Liu, and C. Ratti. Estimating Origin-Destination Flows Using Mobile Phone Location Data. *IEEE Pervasive Computing*, Vol. 10, No. 4, 2010, pp. 36–44.
11. Alexander, L., S. Jiang, M. Murga, and M. C. González. Origin–Destination Trips by Purpose and Time of Day Inferred from Mobile Phone Data. *Transportation Research Part C: Emerging Technologies*, Vol. 58, 2015, pp. 240–250.
12. Li, H. Using Mobile Phone Data to Analyze Origin-Destination Travel Flow Dynamics for City of Pasadena, CA and Surrounding Area. Presented at 94th Annual Meeting of the Transportation Research Board, Washington, D.C., 2015.
13. Gonzalez, M. C., C.A. Hidalgo, and A. L. Barabasi. Understanding Individual Human Mobility Patterns. *Nature*, Vol. 453, No. 7196, 2008, pp. 779–782.
14. Jiang, S., J. Ferreira, and M. C. González. Activity-Based Human Mobility Patterns Inferred from Mobile Phone Data: A Case Study of Singapore. *IEEE Transactions on Big Data*, Vol. 3, No. 2, 2017, pp. 208–219.
15. Dong, H., K. Liu, Y. Zhenyu, Z. Li, L. Jia, Y. Qin, and Y. Wang. Understanding Spatial and Temporal Patterns of

Urban Travel Demands with Call Detail Record Data. Presented at 96th Annual Meeting of the Transportation Research Board, Washington, D.C., 2017.
16. Sun, Z., D. Chen, and X. Li. Identification of Transportation Corridor Based on OD Distribution. *Journal of Transportation Engineering and Information*, Vol. 8, No. 3, 2010, pp. 116–123.
17. Fan, D., and T. Yang. Study of Two Step Gather Method of Urban Main Transportation Flow. *China Journal of Highway and Transport*, Vol. 10, No. 4, 1997, pp. 84–89.
18. Chen, B., Q. Wu, A. Dong, Q. Sun, and F. Ma. Eigen Line Identification for Regional Transportation Corridor Based-on OD Cluster Analysis. *Journal of Transportation Systems Engineering and Information Technology*, Vol. 14, No. 4, 2014, pp. 30–36.
19. Wang, Y., L. Li, J. Yu, Z. Li, S. Wang, and R. Ke. Identifying the Urban Transportation Corridor Based on Mobile Phone Data. *Proc., IEEE 1st International Smart Cities Conference (ISC2)*, 2015. https://dx.doi.org/10.1109/ISC2.2015.7366203.
20. Zhong, G., T. Yin, J. Zhang, X. Chen, D. Yu, and B. Ran. Identifying Passenger Distribution Corridor for the Transportation Hub Based on Mobile Phone Data. Presented at IEEE 20th International Conference on Intelligent Transportation Systems (ITSC), Yokohama, Japan, October 2017.
21. Shanghai Hongqiao Central Business District. *Passenger Flow Information of Hongqiao Hub in 2013*. http://www.shhqcbd.gov.cn/html/shhq/shhq_2013/Info/Detail_6403.htm. Accessed May 15, 2015.
22. Zhong, G., X. Wan, J. Zhang, T. Yin, and B. Ran. Characterizing Passenger Flow for a Transportation Hub Based on Mobile Phone Data. *IEEE Transactions on Intelligent Transportation Systems*, Vol. 18, No. 6, 2017, pp. 1507–1518.
23. Chen, C., L. Bian, and J. Ma. From Traces to Trajectories: How Well Can We Guess Activity Locations from Mobile Phone Traces? *Transportation Research Part C: Emerging Technologies*, Vol. 46, 2014, pp. 326–337.
24. Shanghai City Comprehensive Transportation Planning Institute. *Report of the fourth comprehensive travel survey in Shanghai*. Shanghai, 2010.
25. Zhang, R., and L. Zhu. Impact of Adjustment of Huhang High-speed Railway Departure Stations in Railway Hub on Attraction of Passengers with Different Features. *Journal of the China Railway Society*, Vol. 38, No. 3, 2016, pp. 9-17.

The Standing Committee on Effects of Information and Communication Technologies (ICT) on Travel Choices (ADB20) peer-reviewed this paper (18-01777).

Article

TRR
JOURNAL OF THE TRANSPORTATION RESEARCH BOARD

Transportation Research Record
2018, Vol. 2672(50) 35–45

DOI: 10.1177/0361198118776810
journals.sagepub.com/home/trr
SAGE

Virtual Immersive Reality for Stated Preference Travel Behavior Experiments: A Case Study of Autonomous Vehicles on Urban Roads

Bilal Farooq[1], Elisabetta Cherchi[2], and Anae Sobhani[3]

Abstract

Stated preference experiments have been criticized for lack of realism. This issue is particularly visible when the scenario does not have a well understood prior reference, as in the case of research into demand for autonomous vehicles. The paper presents Virtual Immersive Reality Environment (VIRE), which is capable of developing highly realistic, immersive, and interactive choice scenarios. We demonstrate the use of VIRE in researching pedestrian preferences related to autonomous vehicles and associated infrastructure changes on urban streets in Montréal, Canada. The results are compared with predominantly used approaches: text-only and visual aid. We show that VIRE results in respondents having better understanding of the scenario and it yields more consistent results.

Stated preference (SP) experiments involve presenting the participants with carefully created hypothetical scenarios in order to assess their choice preferences efficiently (*1*). SP experiments are considered to be a powerful tool used in transportation research to evaluate behavioral responses—especially in cases where the alternatives do not exist in the current market. However, they have been criticized for lack of realism in that they do not realistically represent the actual choice situation (*2*). This is particularly true in the case of highly innovative services and technologies, of which respondents have never heard or have very little understanding.

Connected and autonomous vehicles (CAV) are expected to become part of the urban vehicular fleet in the coming years. CAV technology has the potential to revolutionize mobility and to have profoundly effects on land use, regional economics, individual behavior, culture, and society (*3*). Recently, web-based SP survey techniques have been used to study questions like demand for CAV, willingness to pay (WTP), and adoption rates (*3*, *4*). However, these studies primarily rely on the respondents' mental images of CAV and their perception of what CAV may or may not be able to do. Respondents' understanding of the CAV technology and capabilities may be limited and heterogeneous. The lack of experience and hence of knowledge of the product may have a major impact on the estimation of individuals' preferences. The degree of realism possible in a text-heavy web-based SP survey is also limited. Furthermore, for more detailed questions, like understanding pedestrians' perceptions, the lack of actual experience in web-based technologies raises questions about the relevance and quality of their generated datasets.

The use of visualization and simulation in SP experiments has been shown to improve realism (*2*). Patterson et al. concluded that the use of visuals is highly beneficial when they convey accurate descriptions of the choice scenarios. Due to recent technological development of the head mounted display (HMD), virtual and augmented reality is becoming more realistic, representative of reality, and accessible. Various consumer HMDs are now available, including Oculus Rift, HTC Vive, and Microsoft HoloLens. This creates a great opportunity for the use of virtual and augmented reality to develop realistic and representative SP experiments—especially in situations where the respondent's mental image of a technology or service is very limited and it is complicated to express the tasks textually.

[1]Laboratory of Innovations in Transportation (LITrans), Ryerson University, Toronto, ON, Canada
[2]School of Civil Engineering and Geosciences, Newcastle University, Newcastle
[3]Transport and Logistics Group, TU Delft, Delft, the Netherlands

Corresponding Author:
Address correspondence to Bilal Farooq: bilal.farooq@ryerson.ca

This paper presents the Virtual Immersive Reality Environment (VIRE) platform for developing a range of transportation SP experiments that are highly realistic, immersive, interactive, and have the capability to collect additional information about the motion, orientation, and oral reactions ("Oh!"s and "Ah"s) of the respondents. VIRE not only addresses the realism issue in SP experiments, but it also offers the ability to design more detailed experiments that can investigate dimensions like mental processes in choice experiments, without putting any extra burden on the respondent. The information provided to the respondent can also be well controlled and manipulated. We demonstrate the capabilities of VIRE in a choice situation where the preferences of pedestrians in the context of autonomous vehicles on urban roads are investigated. In this controlled experiment, respondents physically cross a virtual road intersection and interact with vehicles. We also compared the VIRE based SP experiment with text-only and standard visualization based SP experiments.

The rest of the paper is organized as follows. The next section presents a brief overview on the use of visualization in SP experiments and the general use of SP experiments in studying the behavioral dimension of the forthcoming technologies. We then present the VIRE platform and its features. The experiment design is presented, which involves the pedestrian walking in a virtual intersection with manual and autonomous vehicles. We then present the results and their analysis. The paper concludes with remarks on the capabilities and possible uses of the VIRE.

Background

The SP experiments research community proactively explores new tools in order to make choice tasks and thus responses more realistic. Patterson et al. (*2*) conducted location choice experiments using text and gaming engine-based animations. They concluded that the combination of visual and textual information increased the processing of information by the respondents. Cherchi and Hensher (*1*) pointed out the need for visual and engaging tools, for example, eye tracking, virtual reality, and simulators, to add value to behavioral relevance and reduce hypothetical bias. However, due to the amount of work needed to actually develop these tools, very little progress has been made in terms of the mainstream usage of more immersive tools in SP experiments.

This fact is evident from the recent literature on the investigation of the effects of CAV technology. For instance, (*5*) and (*4*) used web-based survey tools to determine WTP for automated vehicles among people in the U.S.A. An important dimension of disruptive technology like automated vehicles is that they do not have a strong reference in our daily lives for respondents to associate with. Moreover, most people have not experienced such technology in any form yet. A web-based survey does not have the required features for a SP researcher to design a choice scenario that best represents the reality and also helps the respondents to completely understand the scenarios. It is further evident from the high variance in the WTP reported in two studies that the web-based survey may not have provided the level of information necessary to respond objectively (*5*, *4*). Daniel McFadden, in his keynote speech at the 2015 conference of the International Association of Travel Behaviour Research (IATBR), alluded to this shortcoming in the context of disruptive technologies and suggested the use of immersive simulators that can give respondents a near-real experience.

In recent years there have been rapid advances in improved vehicle automation technology, and several technology providers and auto manufacturers plan to introduce an automated self-driving vehicle by or before 2020 (*6*). These developments will have significant impacts on the transportation system such as roadway safety, traffic congestion, environmental impacts, personal mobility, and travel behavior in general (*7*). CAV can offer greater safety and efficiency within the transportation system and potentially reduce the high proportion (90%) of crashes that result from driver error (*8*).

With advances in vehicle technology over the past few decades, the future of automobile travel has become harder to predict. That said, although there are many researchers studying hybrid-electric vehicles, plug-in electric vehicles, car-sharing services, and on-demand services (Uber, Lyft, etc.), the real demand for the emerging CAV is still under investigation. Recent studies have initiated the evaluation of the real impact of CAV on reduction of crash rates, safety and security, economic effects, etc. (*9*, *10*). However, understanding the future of this emerging mode of transportation is still a real challenge (*5*, *10*), and one of the reasons is precisely the lack of real settings and real experience that would allow individuals to form preferences for CAV and researchers to measure those preferences.

Over the years, several researchers, private enterprises, and industry affiliations have attempted to forecast CAV demand, and have made estimated predictions of CAV adoption rates (*11–15*). However, the predictions are based on extrapolating trends from older methods (e.g. conventional vehicle technologies, expert opinions, or forecasts of supply-side variables) which involve an assumption of consistency between the two different vehicle types, and therefore is not a reliable method given the different nature of CAVs. More reliable demand forecasting requires understanding of CAV-specific factors such as WTP, government regulation of technology adoption, and the effects of different levels of CAV (e.g. level 1 and 2 automation technologies) such as lane centering assistance and adaptive cruise control. Litman (*13*) made use of data on the deployment and adoption of previous smart vehicle technologies (e.g. automatic navigation and hybrid-electric drive) and predicted that, by 2040, 50% of vehicle sales, and 40% of all vehicles, will be CAV. Another forecasting study by Laslau et al. (*12*) predicted that, by 2030, the market size for level 2 and 3 automation technologies will account for up to $87 billion. However, they argue that level 4 technology is likely to be emerging by that time and level 3 automation will still be a premium option, which is expected to account for just 8% of new car sales. Some predictions have gone so far as to forecast level 4 CAV usage to be 90% of all trips. However, the experience with a seemingly new technology (all predictions of the electric vehicles markets proved to be wrong) should be a warning about the risks associated with predictions of the market penetration of technologies that are still in evolution and that people have no experience with. To understand individuals' opinions about CAV, Casley et al. (*16*) conducted a survey of 467 respondents, finding that 30% of them were willing to spend more than $5000 to adopt full automation in their next vehicle purchase and around the same proportion showed interest in adopting autonomous vehicle technology, four years after its introduction on the market. In that study, 82% of respondents ranked safety as the most important factor affecting their adoption of CAV, 12% and 6% ranked legislation and cost respectively as their main reason for choosing CAV. Another survey, by Underwood, considered the opinions of 217 experts. It found that legal liability was the most difficult barrier to the wide implementation of level 5 CAV, and consumer acceptance the least. Additionally, approximately 72% of the experts suggested that CAV should be at least twice as safe as conventional vehicles before they are authorized for public use, while 55% of the experts indicated that level 3 CAVs are not practical because drivers may not take required actions in time (*17*). Schoettle and Sivak found that females are more interested in autonomous vehicle technology than males, who were more concerned about being forced to follow speed limits (*10*). Focusing on different age groups, older participants (60 years and older) and younger participants (21–34 years) expressed the highest WTP in order to obtain self-driving technologies. As an important step in planning, demand estimation and analysis needs to be developed to evaluate the effect of the large-scale arrival of CAV in the car market. The first step to any demand modeling is data collection, which in this case faces several problems due to the very low number of CAV users. To address this issue, studies have mainly used web-based questionnaires about individuals' opinions and preferences toward the adoption of these emerging vehicles (*5*).

Daziano et al. used SP surveys where the level of automation was one of the attributes that described the hypothetical choice situations (*4*). As discussed above, while powerful and necessary in the case of new contexts, SP surveys have been questioned for their lack of realism. The use of visual attributes (images) has been tested in the literature as a tool to improve realism (see recent work by Patterson et al. (*2*)). Visualizations also play an important role in communicating transportation plans and policies. However, the use of visual images in a standard SP setting does not solve another major problem associated with using SP to study preferences for CAVs, that is, the lack of experience and hence knowledge of the product, which has a major impact on the estimation of individuals' preferences (*18*, *19*). Recent advances in computer graphics and technology have provided new opportunities for generating more realistic virtual scenarios that are suitable for behavioral studies (*2*). An emerging interactive technology in the field of transportation studies is virtual reality (VR) tools which have opened a new window for practical applications and scientific investigation of human perception and behavior. VR allows the user to be immersed in a set environment for in-depth evaluation of user perception and behavior. Its ability to overlay the physical environment with virtual elements, such as information or images, which can interact with the physical environment in real time, provides new possibilities for content delivery. VR allows the sensation of immersion in the activities on the screen and the virtual elements (*20–23*). The approach proposed in this paper consists in conducting an SP experiment within a VR environment. This approach is better than a typical SP experiment in terms of realism, because it puts respondents in a realistic environment and allows them to interact with and move within the environment as they would do in real life. Given the role played by realism in SP experiments, the level of realism achieved with a VR environment represents a significant improvement. Additionally, our SP experiment implemented within a VR environment allows respondents to gain direct experience with the product, which is the biggest single drawback of using standard SP studies to measure preferences for CAV. VR environment experiments have been conducted successfully in various fields of cognitive studies. For example, studies have shown that people can develop realistic spatial knowledge in the VR environment that is similar to actual physical environments (*24–26*). The main advantage of adopting VR in research studies is the freedom and versatility in setting up experiments which enables scientists to measure the physical reactions of

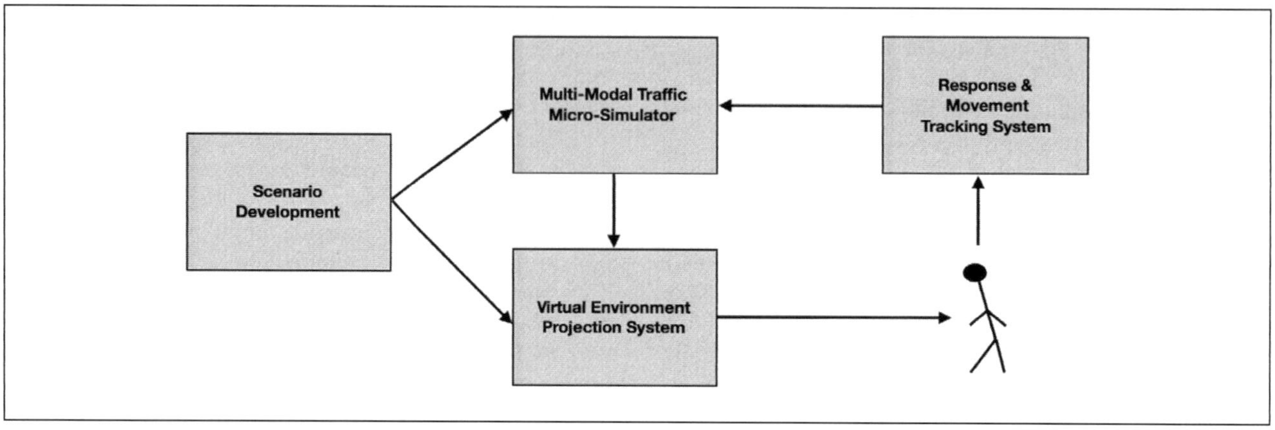

Figure 1. Components and interactions in VIRE.

participants by means of electrocardiography, skin conductance, and electroencephalogy. Eye tracking enables researchers to understand visual perception better by monitoring gaze behavior (*2*, *27*).

A major benefit of using a VR environment is the ability to examine various scenarios for incidents, employ appropriate treatments/policies, and analyse their performances before making irreversible decisions that are either difficult or expensive in the real word. An immersive head mounted virtual reality environment (IHMVR) is a VR display device that uses an optical system to directly present virtual scenes received by the display and works with the human brain to produce a strong sense of immersion (*28*). Immersive virtual environments which allow a locomotive interface may preserve the perception–action coupling that is critical in examining many visual timing skills (*29*). Researchers have also applied virtual environments to study pedestrian route choice and reaction to information in evacuation scenarios. Sobhani et al. studied the effects of multitasking (walking and using smartphone) on adults' street crossing performance by using a simulated intersection crossing task programmed in an IHMVR with an integrated audio and sensors which allowed close emulation of real word behavior of pedestrians (*30*). Using a high-fidelity street crossing simulator, Neider et al. showed that naturalistic cell-phone conversations impair crossing performance and increase crash rates (*31*). Similarly, Stavrinos et al. demonstrated significant costs to simulated crossing performance while conversing on a cell phone (*32*, *33*).

Virtual Immersive Reality Environment (VIRE)

Leveraging upon the technological developments in HMD and VR, VIRE aims to bring realism, interactivity, and immersiveness into the choice scenario. Furthermore, VIRE enables us to design complex experiments to study phenomena like mental processes directly, without putting extra burden on the respondents. In the standard setting, the respondents can experience forthcoming transportation technologies and services very realistically, by projection of the traffic simulation directly to their eyes and real-time interactions between them and simulation objects (e.g. CAV, cyclists, etc.). This also improves the understanding of the scenario by respondents in comparison to conventional techniques like text or visuals-based SP experiments. As shown in Figure 1, VIRE has four main systems that interact with each other and the respondents. These systems are a mix of software and hardware components working together.

Scenario Development (SD) System

The theoretical experiment design is translated into a 3D scenario using Unity, an open-source gaming engine (*34*). The scenario is created using 3D models of the buildings, vehicles, persons, cyclists, trees, roads, and other components necessary. This system is capable of importing already available 3D resources and incorporating them into the scenario. The road network internally is coded as a collection of directed links and nodes. VIRE is capable of importing any road network from the freely available OpenStreetMaps to the VIRE road network.

Multi-Modal Traffic Micro-Simulation (MMTM) System

VIRE currently supports the large-scale simulation of vehicles, person, and cyclists. Vehicular movement is simulated using the Intelligent Driver Model (IDM), which is a variant of a widely used car-following model (*35*). IDM allows us to incorporate the behavior of CAV, especially avoiding collisions with pedestrians and maintaining a safe stopping distance from them. In the VIRE implementation, we used the IDM parameters calibrated by Kenting and Treiber (*35*). For simulating the pedestrian movement, we adapted the social force model (*36*) to incorporate the multi-modal nature of the simulation. Pedestrians not only calculate the forces with respect to other pedestrians and obstacles, but also approaching vehicles and cyclists. In the VIRE implementation, we used the social force parameters calibrated by Sahaleh et al. (*36*). For cyclists' movement, to the best of our knowledge no dedicated flow model has yet been developed. Thus, we adapted the social force model to mimic the cyclists' movement. Note that the currently implemented models and their parameters within VIRE can be changed very easily to meet the specific requirements of an experiment. The simulator is implemented in $C^{\#}$ language in order to interact seamlessly with Unity, which is also implemented in the same development environment. This system uses the network, resources, and environment developed by the SD system to simulate the multi-modal traffic flow.

Virtual Environment Projection System

Using the 3D visualization and animation support available in the Unity gaming engine, this system projects the positions and interactions of various dynamic objects (i.e., vehicles, pedestrians, and cyclists) onto a commercially available HMD. Currently, we are only using Oculus Rift for projection, but our projection system can support other HMDs including HTC Vive, Microsoft HoloLens, and others. The projection system takes the environment with static 3D objects that is created by SD system and overlays the dynamic moving objects, whose positions are provided by the MMTM system. VIRE is also capable of generating and associating audios to 3D objects, for example, a car passing by will make the appropriate noise that the respondent can hear. High importance is given to the scene refresh rate in order to create a realistic environment. Current HMDs are capable of refreshing at 90Hz. Thus, we designed our traffic simulation and scene rendering in such a way that more than 10,000 agents can be simulated while maintaining at least 90Hz. This is achieved by employing parallel processing and localized rendering. We are currently using the Core i7 6700K (4.00GHz, 8M Cache, Quad-Core) processor and Nvidia GeForce GTX 1080 graphics card to ensure maximum processing capabilities.

Response Tracking System

The distinctive feature of VIRE is that it can not only capture various types of responses from the respondent but can also let them interact with other virtual objects (Figure 2). Currently, we can record the exact trajectory of the respondent moving within an area of 4.5 m $\times$ 2 m using the Oculus Motion Sensors. This area can be extended if required, by using appropriate hardware with minimal changes in the core software. VIRE can also record any vocal responses (i.e., the "Oh!"s and "Ah"s). The actual movement of the respondent can be videorecorded. Moreover, VIRE also records the direction the respondent is looking by recording the orientation of their head. All this information about the respondent is transmitted to the MMTM system, which updates the simulation according to the responses and interactions with virtual objects. Additionally, we have implemented a wrapper around Oculus Touch hardware, so that it can be converted into a hand-held device. This can be useful to examine respondents' behavior in terms of smartphone use, driving, etc. This functionality can be extended to any other hand-held device. Sobhani et al. (*30*) is an example of the use of this feature where we studied the behavior of smartphone-distracted pedestrians crossing the road.

Experiment Design

To demonstrate the distinctive functionalities of VIRE system, a controlled SP-like experiment related to the preference for autonomous vehicles in a dense urban environment was designed. An area dominated by pedestrians outside Laurier metro station in Montréal was selected (Figure 3). The intersection of Laurier and Rivard Streets was created in VIRE. The controlled experiment involved two scenarios:

(a) Current state: signalized intersection with vehicles driven by people (Figure 3*a*).
(b) Hypothetical scenario: unsignalized intersection with autonomous vehicles and pedestrian priority.

Both scenarios were developed in VIRE. In the case of scenario (a), pedestrians would wait for the walk signal to turn green. They could jay-walk but were not allowed to run. Running was prohibited because the respondent may hit some physical obstacle in the laboratory or trip themselves with the wires. Furthermore, if they jay-walk and a vehicle is coming, the safe distance may not be maintained. In that case, the respondent would be notified that an accident has happened. In scenario (b), autonomous vehicles would stop at a safe distance if the respondent tried to enter the crossing unsafely. Here too, the respondents were not allowed to run. Based on these two scenarios, we investigated the following questions:

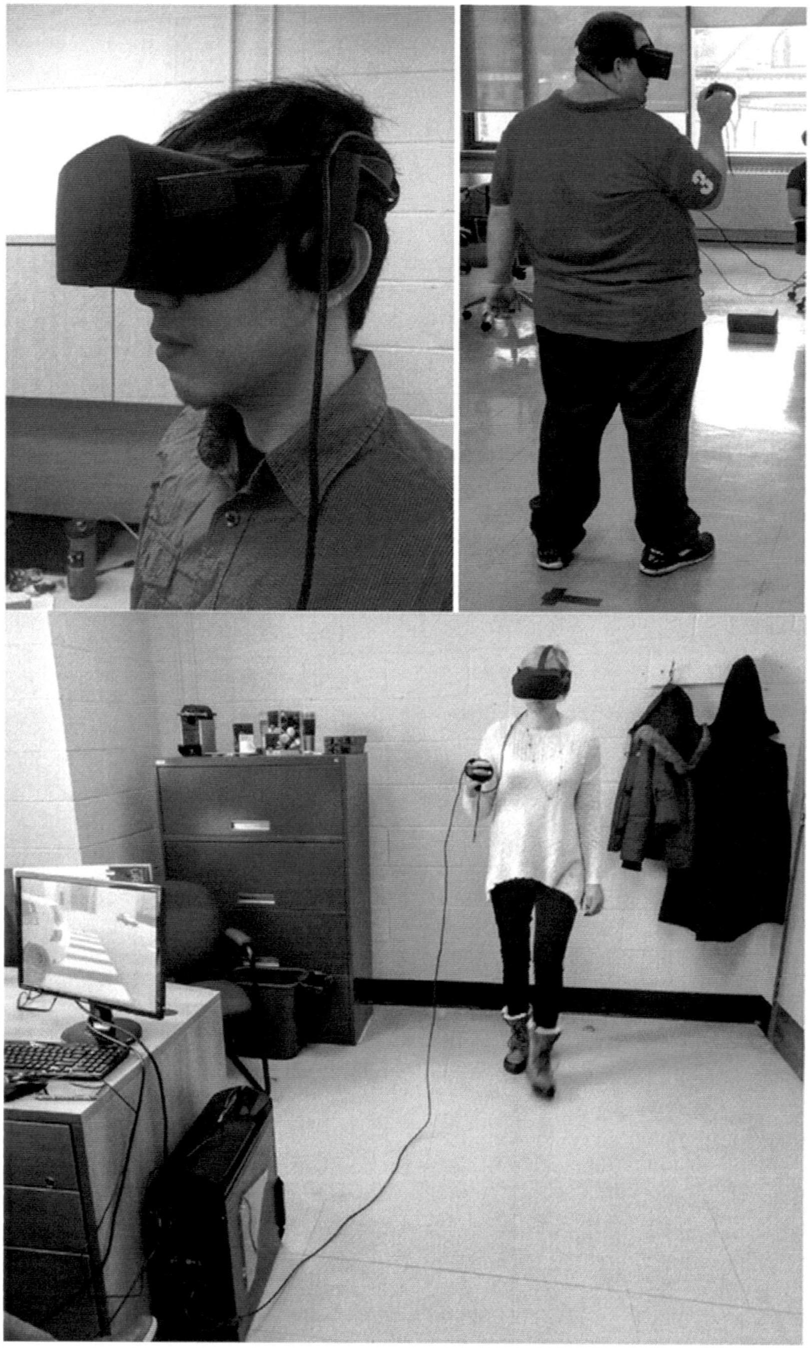

Figure 2. Respondents experiencing the VIRE system.

(a) Analysis of the pedestrian acceptance of autonomous vehicles and associated infrastructure changes in urban settings using VIRE
(b) Comparison of VIRE with existing experiment tools

- Text-only
- Visuals

Respondents were first given a piece of paper that described the two scenarios using text only. They were

Figure 3. Intersection setup for the experiment: (*a*) Google satellite view; (*b*) virtual representation of the intersection.

then asked to choose their preference out of the two. In the second part of the experiment, the respondents were shown an animation of the two scenarios. The animation was repeated several times for each respondent to understand what is going on and to be familiar with the scenario. They were asked again for their preference. In the third part of the experiment, they were given the HMD to wear and physically walked through the two scenarios developed in VIRE. This part of the experiment was repeated 10 times with different arrival times and gaps between the vehicles, which were precomputed. Each respondent would go through the 10 trials in the same order. The pattern was controlled by a Poisson arrival distribution (μ = 4 sec). The horizon of each trial was 2 min. If the respondent was not able to cross during this time, the trial would stop. The speed limit was set at 50 km/h. In each trial, based on safety literature, there were two "safe gaps" randomly inserted between the cars—one was 5 sec and other was 7 sec. This gap was such that if the pedestrian crossed, the incoming car would not need to stop. The respondent's trajectory in the virtual environment, head movement, and vocal and physical expressions were also recorded. During the experiments, respondents were constantly asked if they felt uncomfortable or experienced symptoms of motion sickness. We were ready to stop the experiment immediately if the respondent replied positively. However, none of the participants experienced motion sickness.

In total 42 respondents in Montréal and Toronto participated in these experiments. Our sample included university students older than 18 years, both male and female. Basic socio-demographics and information on their dominant mode and activity patterns were collected. Their level of familiarity with VR was also recorded. In terms of mode of transport, 12% reported that their primary mode was biking, 45% public transit, 19% walk, and 24% car.

Results and Analysis

Among the respondents, 43% were female and 57% male. The average age of the respondents was 26 years, with minimum of 19 years, maximum 40 years, and a standard deviation of 5 years. The respondents were residents of Toronto (41.5%) and Montréal (58.5%), Canada; 80% of the respondents had a driving license; and 36% of the respondents reported that they had experienced the HMD-based VR before.

Figure 4 shows the comparative analysis of the choice preferences using the three experiment tools: text-only,

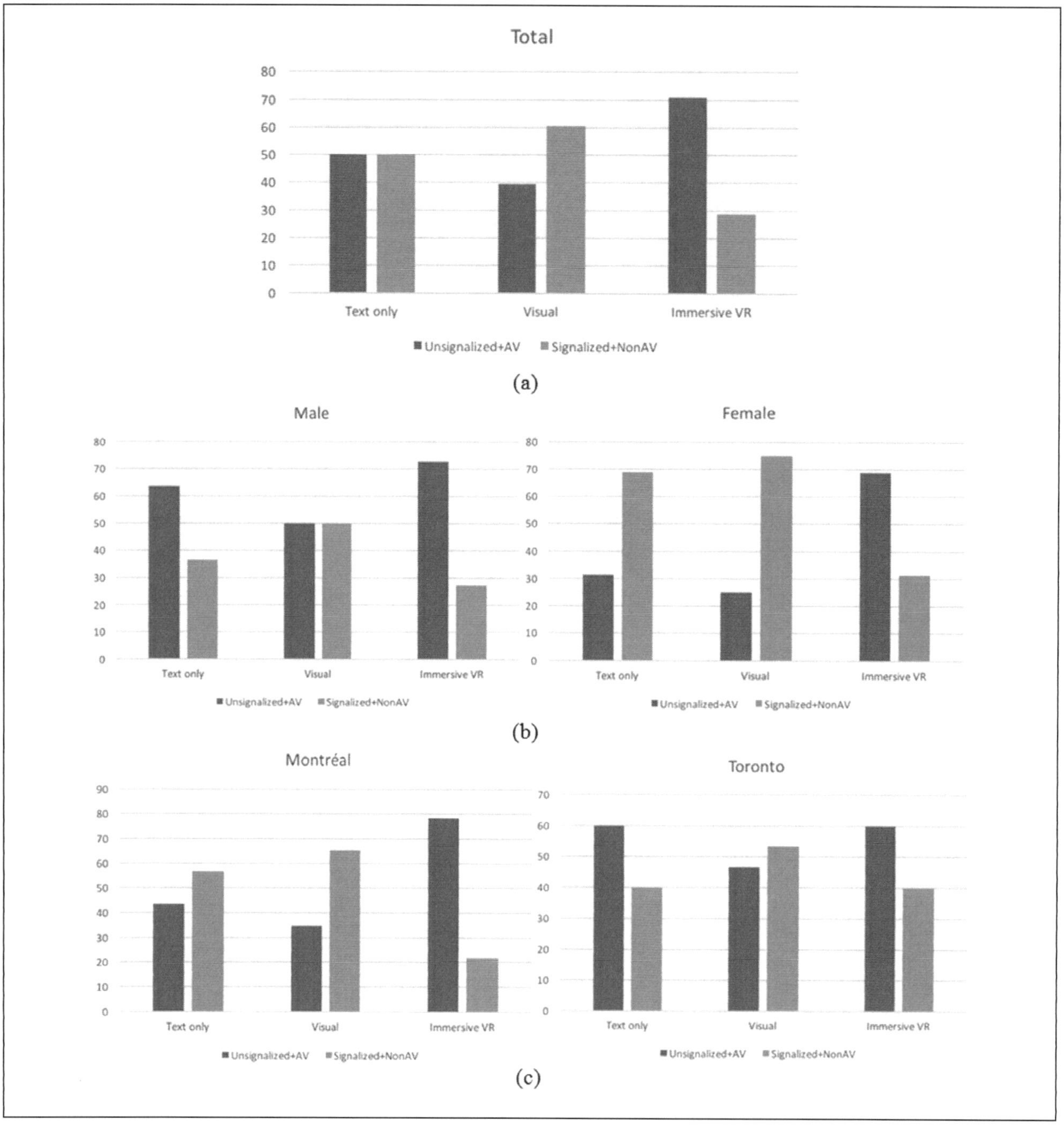

Figure 4. Choice comparison between text-only, visual, and VIRE: (*a*) overall choice behavior; (*b*) choice behavior by gender; (*c*) choice comparison by city.

visual, and VIRE. In the case of text-only (Figure 4*a*), 50% of the respondents reported that they would prefer the scenario where autonomous vehicles are operating on urban streets without signalization and with pedestrian priority. In the case of simple visualization this percentage decreased below 40%. We hypothesize that the reason for this decrease in preference is that the information provided by the second tool (visual animation) does not add any value in terms of scenario realism. Instead, it affects the respondent's perception of autonomous vehicles in a negative way. Using VIRE, the preference for the autonomous vehicle scenario increased to 70%.

Table 1. Multinomial Logit Model Estimates

Parameters	Text		Visualization		VIRE	
	Value	Std error	Value	Std error	Value	Std error
ASC_{CAV}	1.90	1.87	−2.12	−1.88	−3.49	−1.64
β_{Age}	0.69	0.69	0.65	0.63	1.76	0.83
β_{Female}	−1.61	0.80	−1.42	0.82	−1.14	−1.32
$\beta_{Bike, Male}$	0.972	1.72	0.21	1.12	12.70	0.92
$\beta_{Toronto}$	−1.09	0.75	−0.82	−0.74	−1.18	−0.82
β_{HMD}	–	–	–	–	2.03	0.84
$\mathcal{L}$	−25.28		−25.18		−18.98	
ρ^2	0.132		0.135		0.348	
Observations	42		42		42	

In this case, the realism of the scenario is improved, resulting in a near-realistic, immersive, and interactive experience for respondents.

Clear differences between the genders are noticeable (Figure 4*b*), with female respondents being more conservative about the autonomous vehicles than males in all three experimental tools. In the case of all three tools, males are more likely to prefer the scenario with autonomous vehicles, and this preference increases when using the VIRE. Female respondents prefer (about 70%) the current state (i.e., normal vehicles and signalized intersection) when the experiment is run with text-only or simple visualization. However, this percentage flips when the female respondents experience the scenario in VIRE. There are some notable differences between respondents from Toronto and Montréal (Figure 4*c*) as well. In general, respondents in Toronto were more conservative about autonomous vehicles while using VIRE as the experiment tool. This can be attributed to the general differences between the urban streets of Toronto and Montréal. Toronto is a typical North-American city with wider roads, higher volume of traffic and speed limits, and longer block-sizes. Central Montréal has a more European feel, with narrower roads, lower speed limits, smaller blocks, and more mixed-use areas. However, it should be noted that due to the small sample size, we cannot generalize this observation.

To assess quantitatively the factors affecting the respondents' preferences and differences between the tools used, we developed three multinomial Logit models (MNL). Table 1 reports the three models with parameter values that were found to be statistically significant. We also jointly estimated the three models by fixing the scale parameter ($\mu = 1$) for the VIRE model and estimated the scale for the other two. The estimated scale parameter for MNL based on text-only and basic visualization was found to be 1.16, which indicates differences in the samples. The model developed from VIRE has better overall statistics with lowest log-likelihood. All the individual parameters in VIRE based MNL had right expected signs. In terms of parameter values, the older respondents (millennials) had a greater preference for the autonomous vehicles scenario. Compared with males, females had a lower preference for autonomous vehicles. Males whose primary mode of travel was biking preferred the autonomous vehicle scenario. This may be due to the fact that they expect autonomous vehicles always to follow the rules and thus they feel safer. Respondents from Toronto had a lower preference for autonomous vehicles compared with those from Montréal. Furthermore, prior experience with HMD had a positive and significant effect on the choice of the autonomous vehicle scenario. This may be due to the fact that in general such respondents felt more comfortable using the HMD and were able to absorb more information in the scene.

Concluding Remarks

We present Virtual Immersive Reality Environment (VIRE), a generic platform for developing travel related SP experiments that are highly realistic, interactive, immersive, and can efficiently collect information beyond just preferences. We demonstrate the use of VIRE in the preferences of pedestrians related to autonomous vehicles and the related infrastructure changes. Pedestrians can physically walk in a highly realistic virtual 3D model of a real intersection in Montréal. The data collected using VIRE is compared with the predominantly used tools: text-based and visual animation. We found that the understanding of respondents increased, and their preferences became more consistent. The joint MNL model estimated also showed the difference in scale parameter for VIRE and other tools. In the VIRE experiment we controlled for the arrival and gap of the vehicles. We plan to include that in the model to understand the effects of level of autonomous vehicle traffic and gaps on the pedestrian preferences.

SP experiments have been known to suffer from lack of realism. This issue becomes even more relevant when the transportation technology or service is highly innovative and without a well understood reference, for example, autonomous vehicles and their operations on urban roads. VIRE not only addresses the realism issues but also goes beyond them by collecting additional information on the movement, vocal responses (i.e., the Ohs and Ahs), the direction in which the respondent is looking, and interactions with other objects, without any additional burden on the respondent. This information can aid in developing more complex and realistic models of behavior. One such application could be the explicit modeling of respondents' mental processes while deciding on choice preference. Our next step is to design and develop SP experiments in VIRE that can allow us to investigate this line of research.

In terms of functionality, we are currently working on incorporating manual driving hardware (steering wheel, brake, and acceleration) in VIRE, in order to develop experiments that compare the experience of driving manually versus being driven in autonomous vehicles. In the near future, we plan to enhance VIRE so that it can support multiple respondents in one experiment. This functionality would be very useful in studying group behavior, for example, two friends walking together or a family driving in an autonomous vehicle.

One of the biggest hurdles in the mainstream use of virtual and augmented reality (VR/AR) in travel related SP experiments has been the amount of software engineering, traffic flow theory, and simulation related knowledge and effort required to develop the data collection tool. In the near future, we plan to bring down this entrance barrier by sharing the VIRE source code with the transportation research community. We envision that this will extensively enhance our collective ability to use VR/AR for understanding travel behavior, its drivers, and the influence of new technology and services. We plan to conduct this experiment on a larger sample involving 300 to 600 participants. A larger dataset will allow us to develop richer models and test our hypotheses more extensively.

Author Contributions

The authors confirm contribution to the paper as follows: study conception and design: Farooq, Cherchi, Sobhani; data collection: Farooq, Sobhani; analysis and interpretation of results: Farooq, Cherchi; draft manuscript preparation: Farooq. All authors reviewed the results and approved the final version of the manuscript.

References

1. Cherchi, E., and D. A. Hensher. Workshop Synthesis: Stated Preference Surveys and Experimental Design, an Audit of the Journey so far and Future Research Perspectives. *Transportation Research Procedia*, Vol. 11, 2015, pp. 154–164.
2. Patterson, Z., J. M. Darbani, A. Rezaei, J. Zacharias, and A. Yazdizadeh. Comparing Text-Only and Virtual Reality Discrete Choice Experiments of Neighbourhood Choice. *Landscape and Urban Planning*, Vol. 157, 2017, pp. 63–74.
3. Fagnant, D. J., and K. M. Kockelman. The Travel and Environmental Implications of Shared Autonomous Vehicles, Using Agent-Based Model Scenarios. *Transportation Research Part C: Emerging Technologies*, Vol. 40, 2014, pp. 1–13.
4. Daziano, R. A., M. Sarrias, and B. Leard. Are Consumers Willing To Pay To Let Cars Drive for Them? Analyzing Response to Autonomous Vehicles. *Transportation Research Part C: Emerging Technologies*, Vol. 78, 2017, pp. 150–164.
5. Bansal, P., K. M. Kockelman, and A. Singh. Assessing Public Opinions of and Interest in New Vehicle Technologies: An Austin Perspective. *Transportation Research Part C: Emerging Technologies*, Vol. 67, 2016, pp. 1–14.
6. Bierstedt, J., A. Gooze, C. Gray, J. Peterman, L. Raykin, and J. Walters. Effects of Next-Generation Vehicles on Travel Demand and Highway Capacity. *FP Think Working Group*, 2014, 10–11.
7. LaMondia, J. J., D. J. Fagnant, H. Qu, J. Barrett, and K. Kockelman. Long-distance Travel Mode Shifts Due to Automated Vehicles: A Statewide Mode-Shift Simulation Experiment and Travel Survey Analysis. Presented at 95th Annual Meeting of the Transportation Research Board, Washington, D.C., 2016.
8. Bellis, E., and J. Page. *National Motor Vehicle Crash Causation Survey (NMVCCS) SAS Analytical Users Manual*. Technical Report. DOT HS 811 053. NHTSA, Washington, D.C., 2008.
9. Fagnant, D. J., and K. Kockelman. Preparing a Nation for Autonomous Vehicles: Opportunities, Barriers and Policy Recommendations. *Transportation Research Part A: Policy and Practice*, Vol. 77, 2015, pp. 167–181.
10. Schoettle, B., and M. Sivak. A Survey of Public Opinion about Autonomous and Self-driving Vehicles in the USA, the UK, and Australia. Technical Report, UMTRI-2014-21, University of Michigan, 2014.
11. Hars, A. *Autonomous Vehicle Roadmap: 2015-2030*. http://www.driverless-future.com/?p=678. 2014. Economic Issues.
12. Laslau, C., M. Holman, M. Saenko, K. See, and Z. Zhang. Set Autopilot for Profits: Capitalizing on the $87 Billion Self-Driving Car Opportunity. *Lux Research*, 2014.
13. Litman, T. Autonomous Vehicle Implementation Predictions. *Victoria Transport Policy Institute*, Vol. 28, 2014.
14. Mosquet, X., T. Dauner, N. Lang, N. Rubmann, A. Mei-Pochtler, R. Agrawal, and F. Schmieg. Revolution in the Driver's Seat: The Road to Autonomous Vehicles. *Boston Consulting Group*, 2015, pp. 11–30.
15. Rowe, R. Self-Driving Cars. *Timeline*, 2015. http://www.topspeed.com/cars/car-news/self-driving-cars-timeline-ar169802.html.

16. Casley, S., A. Jardim, and A. Quartulli. *A Study of Public Acceptance of Autonomous Cars*. Bachelor thesis. Worcester Polytechnic Institute, Worcester, Mass., 2013.
17. Underwood, S. Automated Vehicles Forecast Vehicle Symposium Opinion Survey. Presented at the Automated Vehicles Symposium, San Francisco, Calif., 2014.
18. Jensen, A. F., E. Cherchi, and S. L. Mabit. On the Stability of Preferences and Attitudes Before and After Experiencing an Electric Vehicle. *Transportation Research Part D: Transport and Environment*, Vol. 25, 2013, pp. 24–32.
19. Cherchi, E. *A Stated Choice Experiment To Study the Effect of Social Conformity in the Preference for Electric Vehicles*. Technical Report, Newcastle University, 2016.
20. Animesh, A., A. Pinsonneault, S.-B. Yang, and W. Oh. An Odyssey into Virtual Worlds: Exploring the Impacts of Technological and Spatial Environments on Intention to Purchase Virtual Products. *MIS Quarterly*, Vol, 35, No. 3, 2011, pp. 789–810.
21. Faiola, A., C. Newlon, M. Pfaff, and O. Smyslova. Correlating the Effects of Flow and Telepresence in Virtual Worlds: Enhancing Our Understanding of User Behavior in Game-Based Learning. *Computers in Human Behavior*, Vol. 29, No. 3, 2013, pp. 1113–1121.
22. Jennett, C., A. L. Cox, P. Cairns, S. Dhoparee, A. Epps, T. Tijs, and A. Walton. Measuring and Defining the Experience of Immersion in Games. *International Journal of Human-Computer Studies*, Vol. 66, No. 9, 2008, pp. 641–661.
23. Nah, F. F.-H., B. Eschenbrenner, and D. DeWester. Enhancing Brand Equity Through Flow and Telepresence: A Comparison of 2D and 3D Virtual Worlds. *MIS Quarterly*, 2011, pp. 731–747.
24. O'Neill, M. J. Effects of Familiarity and Plan Complexity on Wayfinding in Simulated Buildings. *Journal of Environmental Psychology*, Vol. 12, No. 4, 1992, pp. 319–327.
25. Ruddle, R., S. Payne, and D. Jones. Navigating Buildings in Virtual Environments. Experimental Investigations Using Extended Navigational Experience. *Journal of Experimental Psychology: Applied*, Vol. 3, No. 2, 1997, pp. 143–159.
26. Tlauka, M., and P. N. Wilson. Orientation-free Representations from Navigation Through a Computer-Simulated Environment. *Environment and Behavior*, Vol. 28, No. 5, 1996, pp. 647–664.
27. Wiener, J. M., C. Hölscher, S. Buächner, and L. Konieczny. Gaze Behaviour during Space Perception and Spatial Decision Making. *Psychological Research*, Vol. 76, No. 6, 2012, pp. 713–729.
28. Wang, Y., W. Liu, X. Meng, H. Fu, D. Zhang, Y. Kang, R. Feng, Z. Wei, X. Zhu, and G. Jiang. Development of an Immersive Virtual Reality Head-Mounted Display with High Performance. *Applied Optics*, Vol. 55, No. 25, 2016, pp. 6969–6977.
29. Wu, H., D. H. Ashmead, and B. Bodenheimer. Using Immersive Virtual Reality To Evaluate Pedestrian Street Crossing Decisions at a Roundabout. *Proc., 6th Symposium on Applied Perception in Graphics and Visualization*, Chania, Crete, ACM, New York, 2009, pp. 35–40.
30. Sobhani, A., B. Farooq, and Z. Zhong. Immersive Head Mounted Virtual Reality Based Safety Analysis of Smartphone Distracted Pedestrians at Street Crossing. Presented at the Road Safety and Simulation Conference, No. 1-12, 2017.
31. Neider, M. B., J. S. McCarley, J. A. Crowell, H. Kaczmarski, and A. F. Kramer. Pedestrians, Vehicles, and Cell Phones. *Accident Analysis & Prevention*, Vol. 42, No. 2, 2010, pp. 589–594.
32. Stavrinos, D., K. W. Byington, and D. C. Schwebel. Effect of Cell Phone Distraction on Pediatric Pedestrian Injury Risk. *Pediatrics*, Vol. 123, No. 2, 2009, pp. 179–185.
33. Stavrinos, D., K. W. Byington, and D. C. Schwebel. Distracted Walking: Cell Phones Increase Injury Risk for College Pedestrians. *Journal of Safety Research*, Vol. 42, No. 2, 2011, pp. 101–107.
34. Unity Developers. *Unity User Manual* (2017.1). https://docs.unity3d.com/Manual/index.html. 2017. Accessed: May 16, 2018.
35. Kesting, A., and M. Treiber. Calibrating Car-following Models by Using Trajectory Data: Methodological Study. *Transportation Research Record: Journal of the Transportation Research Board*, 2008. 2088: 148–156.
36. Sahaleh, S., M. Bierlaire, B. Farooq, A. Danalet, and F. Hänseler. Scenario Analysis of Pedestrian Flow in Public Spaces. Presented at the 12th Swiss Transport Research Conference, No. EPFL-CONF-181239, Lausanne, Switzerland, 2012.

The Standing Committee on Effects of Information and Communication Technologies (ICT) on Travel Choices (ADB20) peer-reviewed this paper (18-02075).

Article

TRR
JOURNAL OF THE TRANSPORTATION RESEARCH BOARD

Transportation Research Record
2018, Vol. 2672(50) 46–55

DOI: 10.1177/0361198118796712
journals.sagepub.com/home/trr
SAGE

Measuring the Effectiveness of a Transit Agency's Social Media Engagement with Travelers

Joanne Douglass[1], Dilum Dissanayake[2], Benjamin Coifman[3], Weijia Chen[4], and Fazilatulaili Ali[2]

Abstract

This study investigated the uses of social media for travel planning on a transit system with particular attention to travel disruptions and delays. As a result of very limited research in the effectiveness of social media in a transit setting, the best practices have yet to be established. Rather than having a one-size-fits all traveler information system, these online services have the potential to provide personalized information tailored to the individual or route they are traveling. Key to this personalization is understanding the audience and their needs. This study sought to explore how, and at what level, transit riders utilize real-time travel information from the social media sites maintained by the transit agency. An online questionnaire was used to collect data about the transit agency's social media users, these data were evaluated using Principal Component Analysis (PCA) and cross-tabulation analysis. Perhaps not surprisingly, there is a significant relationship between respondents' age and their travel purpose. Across age groups and across travel purposes the vast majority of respondents said they most commonly check the transit agency's social media pages, "before journey," to gather daily updates before starting their journeys. Thus, the social media sites already have the potential to influence the users' travel plans before their journey, such as changing their route, travel mode, and/or departure time. On the other hand, this outcome indicates that an active engagement with social media is still missing from the viewpoint of customers. As there are so many users that visit the social media pages for trip planning there is an opportunity to reach these individuals and provide a much more dynamic engagement. While the focus is on a single transit agency, most of the results transcend the specific location or agency.

The Internet and developments in online technology have changed the way we live today, leading to a large shift in the way information can be accessed and an increase in channels open for customer communication. A key reason for this change is the creation of social media and the mass take-up of the technology worldwide (*1*). Access to information is very important for all public services. As a result, public transit agencies are under increasing pressure to engage with their customers who are consuming information beyond traditional media (*2*).

Traditionally transit information was provided in the form of static information such as timetables in leaflets, booklets, or signage at stops and stations (*3*), but this began to rapidly change with the widespread adoption of the Internet, and related information and communication technologies. Transit information is an important resource for passengers to plan their journeys, and many transit agencies view this information as a major factor in attracting people to switch to public transport (*4*, *5*). Today many transit agencies supply transport information tailored to multiple audiences, over a variety of different channels. Generally, passengers require different types of information depending upon the stage of their journey (*5*, *6*). For example, real-time information is often of great value to travelers during disruptions to service (*7*). To reduce traveler frustration and reassure passengers, the disruptions must be managed effectively, and information must be clear and concise for passengers to feel in control (*8–10*). During a disruption passengers have expressed that they would like information quickly;

[1]Nexus, Newcastle upon Tyne, UK
[2]School of Engineering, Newcastle University, Newcastle, UK
[3]Civil, Environmental and Geodetic Engineering; Electrical and Computer Engineering, Ohio State University, Columbus, OH
[4]Transport Development and Planning, AECOM, Newcastle upon Tyne, UK

Corresponding Author:
Address correspondence to Dilum Dissanayake:
dilum.dissanayake@ncl.ac.uk

that is, passengers are favoring real-time information (*11*).

By using social media, transit agencies can benefit from communicating more efficiently with their customer base, thereby saving money and resources when compared with traditional communication methods (*12*, *13*). Many transit agencies in the UK have established a presence on social media networks as a means of disseminating operational information and conversing with passengers to maintain customer satisfaction and resolve complaints. The way that each traveler uses the information available from social media websites will vary across the population; some may use the information only as a source of knowledge and others may consume the information in a useful way for actively planning their journeys. The study presented here used a survey and subsequent analysis to investigate the efficacy of one such social media outreach program by Tyne and Wear Metro (henceforth "TW-Metro"), operator of a light rail transit network, but the results are also of general relevance to other forms of online presence and for transit agencies of all modes. The paper uses TW-Metro customers' engagement and consumption-related attributes together with other socio-demographic and journey-related characteristics. The study investigates how, and at what level, the TW-Metro users utilize real-time travel information from TW-Metro social media pages on Facebook and Twitter.

The paper starts with a critical review of relevant literature. This is followed by details of the data collection process, data analysis, and discussion of the results. The final section summarizes the research findings.

Literature Review

This section provides background to place the work in context. The review covers transit disruptions with an emphasis on communicating information to the travelers, change through technology, and transit agencies' adoption of social media.

Transit Disruption and Information Provision to Customers

Transit disruptions are categorized in the literature as disruption to infrastructure and disruption to the operation of the transport system (*14*, *15*). Disruption to public transit prevents an individual from making their journey from an origin to a destination when or as they had planned, thus having a cascading effect on the individual's planned activities at their destination. Customers want to feel in control of their travel, but the lack of information and the limited understanding of how to interpret the information during a disruption are major factors causing customers to feel frustrated (*8*, *10*, *16*). Ambrosino et al. note that there is a link between travel disruption and passenger information, with the availability of information stimulating confidence about the journey and having the potential to result in setting up alternative plans during disruptions (*17*). Community of Metros explains that information required during a disruption is complex and goes beyond simply announcing that there is a delay (*7*). During a disruption passengers can be affected in different ways depending on their involvement or level of experience (*18*). Different messages will be required depending upon the stage of the individual's journey and involvement in the disruption, as this can affect different travel decisions. For example, potential customers about to begin their trips may acknowledge information regarding their planned journey and take an alternative route or take a different mode, while those directly involved are more likely to desire information regarding a solution (*7*).

Change through Technology

Intelligent Transport Systems have created a major technological shift in transportation (*19*). The Internet without a doubt is the fastest growing information channel, and the wholesale adoption of mobile devices, combined with online communication networks, is rapidly changing the information exchange paradigm once more (*20*). The ability to have an instant connection anytime, anywhere on the go is blurring the lines between people's online and offline lives (*10*). Some 70% of UK adults regularly carry smartphones, and more than half (57%) of all Internet users who ever go online to look at social media sites or apps say they "mostly" do so through a smartphone (*21*).

A social media is defined by Boyd and Ellison as a web-based service that allows individuals to create a public profile, identify a number of consumers and businesses they share a connection with, plus view activity and converse with those people (*22*). Facebook is a social media site and is defined as a social utility that aims to connect people and let them discover what is happening in the world (*23*). As Facebook is a closed network, it is assumed that people are proactive in joining a community within Facebook they are interested in, or in this case a transport service (*10*). Twitter, on the other hand, is an open network, where all content is open to the public and anybody can get involved in the conversation (*24*). The social media sites of Facebook and Twitter are fast growing online outlets that facilitate real-time interaction (*23*). In 2016, Facebook had 1.5 billion monthly active visitors worldwide, and 31 million of those are in the UK; Twitter figures show 15 million monthly active visitors in the UK (*25*).

Adoption of Social Media by Transit Agencies

Transit agencies have sought to keep up with advances in information technology by offering more passenger-centric services (*19*). They are creating official social media pages to update customers about operations and any service disruption throughout the day (*24*, *26*). Baird and Parasnis state that social media networks are where people (i.e., potential transit customers) are congregating, and understandably where businesses want to have a presence (*27*). Clifton states that although transport providers have begun to engage with passengers through social media, some are making better use of it than others (*26*). Clifton also suggested that the better feeds are those that have the right balance of information without overloading the followers' feed; they are manned with real people who seem to know when it is appropriate to joke and when a serious attitude is needed. As a result of very limited research in the effectiveness of social media in transport, from a customer point of view it is very difficult to measure what is best practice. It is well observed that public services have been slow off the blocks when it comes to social media, therefore specific guidance and understanding of the best practices for the sector have also been slow in coming, with many public services deciding to avoid risk that can occur with social media use.

Case Study—Tyne and Wear Metro

The study presented here aims at exploring how riders on the TW-Metro utilized the Twitter and Facebook sites maintained by the transit agency; however, as discussed below, the findings transcend the specific transit agency and the specific social media outlets.

The TW-Metro is a light rail system in North East England; it serves Tyne and Wear region covering its five districts, Newcastle upon Tyne, Gateshead, South Tyneside, North Tyneside, and Sunderland. It was Britain's first light rapid transit system and the heart of an integrated transport network in North East England (*28*); on the other hand, it is considered as the second-largest of the four metro systems in the UK, after the London Underground (the others being the Docklands Light Railway and the Glasgow Subway). The TW-Metro opened to the public in August 1980. Some extensions to the original network were opened in 1991 and 2002; at present the network covers 77.5 km and has two lines with a total of 60 stations, nine of which are underground (*29*). In 2016–17, ridership numbers of TW-Metro users were estimated as 37.2 million passengers per annum (*30*). In the last 10 years, the patronage has fluctuated between 36 million and 41 million passengers.

In the mid-1990s TW-Metro pioneered mobile phone connectivity in its tunnels and was the first railway in Britain to play classical music at stations to improve the passenger waiting environment (*28*). At present, TW-Metro does not have a technology set up to transmit real-time information to the metro users by means of digital and handheld devices. However, there are mobile applications to provide timetable information. Beyond this, the other traveler information services include radio and online media stations which broadcast social media updates, on-station information systems including passenger information displays which have a real-time feed from the track circuits, and passenger announcements which are either automated in real time or can be supplemented by a bespoke announcement from the TW-Metro Control Centre.

According to the NEXUS Public Transport Executive (PTE) which owns and manages TW-Metro, the further development of social media services, for example TW-Metro Twitter and Facebook pages, is their current focus as it may allow to address the gap in lack of delivery of real-time information to some extent while maintaining better and closer interaction with TW-Metro users. The number of updates made every day is predominantly driven by the performance of the system; for example, on a day where there were no disruptions there would only be 2–3 service updates per day. With 450 train services per day, this means that there may be a daily occurrence or an incident ranging from slight delays to major disruptions. Therefore the frequency of information updates on Facebook and Twitter will be highly dependent on the status of the service. As reported by the NEXUS PTE, the growing use of social media by the TW-Metro users has led the organization to invest in a dedicated staff resource to provide updates and answer questions and queries. This is usually one person on duty through the core of the operational day during 07:00–21:00, though staff will work longer hours where there is a need. This focus has increased the number of updates, as the staff is now able to provide more detailed and frequent updates during a major incident. It has also improved the response time, as the staff members on duty are almost always able to respond to a post within 1 hour, but usually much quicker. The NEXUS PTE's view is that they now prioritize Twitter as a medium because the chronology of the user interface means it is much easier for a customer to see the latest information in comparison to Facebook posts.

Data Collection

The survey was developed to collect data in order to investigate the use of social media by TW-Metro passengers for their journey planning process, and also to identify future expansion of communication channels between the transit agency and the customers to provide better customer satisfaction, enhanced trust, and increased demand for the TW-Metro service.

Questionnaire Survey and Data Collection

A comprehensive online survey was designed, giving due attention to the findings from the literature review, and implemented with the intention of collecting an adequate sample for analyzing social media as a communication channel. The survey was published on the TW-Metro social media pages of Facebook and Twitter through the TW-Metro marketing department. The respondents were encouraged to share the survey link with others, and the survey was open for 6 weeks in total. It is fairly common in recent years that questionnaire surveys are posted on social media websites; however, it has been reported that 60% of the questions in questionnaires of this sort are usually unanswered by the respondents (*31*). From the initial set of 450 submitted surveys, 210 were excluded due to missing/incomplete values in at least one of the socio-demographic, journey-related, and social media usage-related categories; yielding, altogether, a final sample size of 240 respondents. The sample represented roughly 1% of TW-Metro social media users at the time of the survey in 2013 (13,247 Facebook followers and 16,027 Twitter followers). It was reported by the NEXUS PTE that number of Twitter followers has increased to 111,035 and Facebook followers to 39,222 since June 2016, after TW-Metro introduced the dedicated staff resource.

The initial part of the survey included general elements of social media use in general and participants' awareness/opinions of the TW-Metro pages in particular. It included frequency of visits to the TW-Metro social media pages and the use of the pages as a channel of communication. The main emphasis was to collect information from users of the TW-Metro social media pages and the majority of the questions were designed to target this group. The final portion of the survey asked about travel habits and demographic information. The survey design incorporated skip-logic to ensure participants were guided through the survey based on the answers to their previous question.

Descriptive Analysis of the Data

The survey responses were analyzed to develop an understanding of the data composition in terms of general social media use, awareness of the TW-Metro social media pages, and visit of these pages. According to the dataset, 93% of the respondents were aware of the TW-Metro social media pages (this high level of awareness makes sense given the method of survey implementation). However, Figure 1 shows that the fraction of respondents

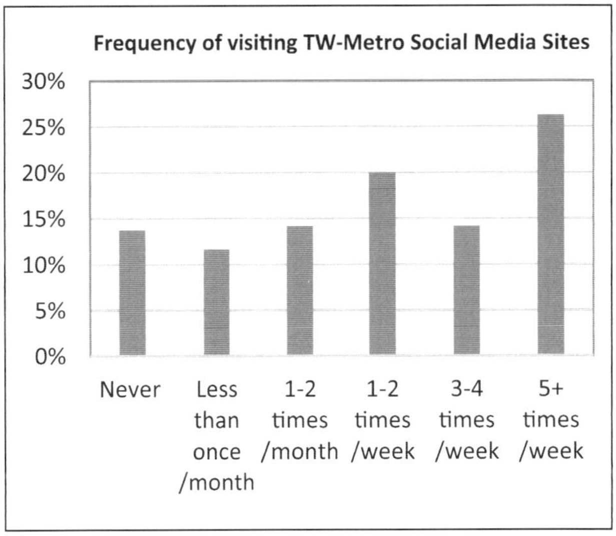

Figure 1. Database statistics.

that visit the TW-Metro social media pages regularly (3 or more times per week) is much lower, at only 40%, with almost 14% of the sample never visiting the TW-Metro social media pages at all. These numbers shift when accounting for riding frequency. Among the riders that travel three or more times per week, 57% check the TW-Metro social media pages at least three times per week; of the remaining responses that reported riding on TW-Metro at all, 52% checked the TW-Metro social media pages at least as frequently as they travel on TW-Metro. Regardless, although these respondents are aware of the social media pages, many do not seem to visit or use them.

Excluding the 14% who never visit the TW-Metro social media pages (and thus, were not asked about the following), the survey asked which single stage is the most common stage of the journey they check the pages for service information. Figure 2*a* shows that the most frequent choice was, *before the start of the journey* (61%), which was followed by, *only if the journey is disrupted* (21%). The majority of respondents found the social media pages most useful for planning their journey, for example, to plan accordingly in the event of any delays or disruptions to service, before leaving home. As the users had to only choose one of five options, it is likely that most will check the social media pages under the other conditions beyond their single most common, but the survey did not ask about such secondary activities.

In terms of demographic profile across all 207 respondents who use the TW-Metro social media pages, Figure 2*b* shows that the two largest groups of respondents were aged 16–24 (37%) and 25–44 (50%), with a smaller number of responses received by 45–64-year-olds (14%). This distribution reflects the tendencies of each

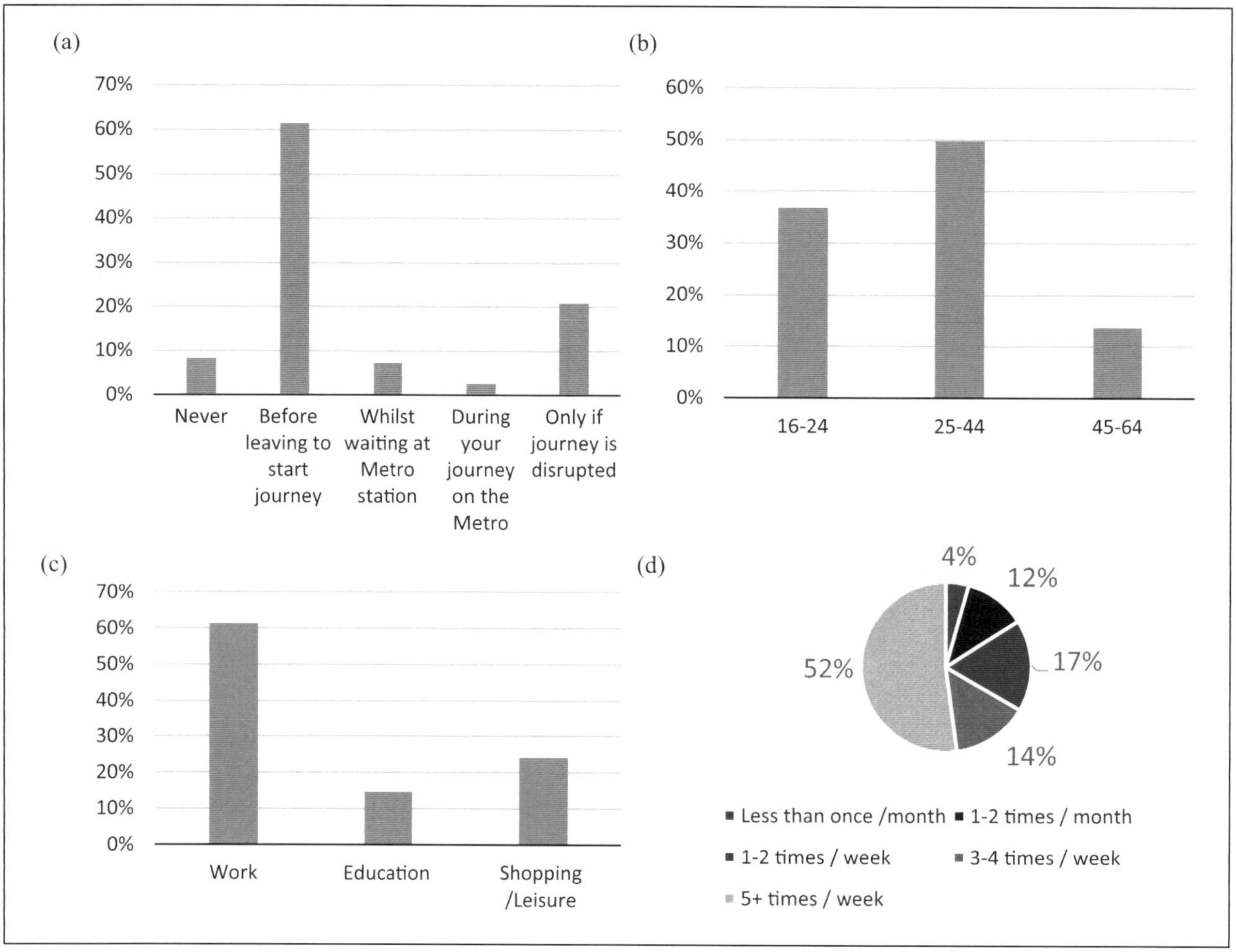

Figure 2. Descriptive statistics of the TW-Metro social media users—percentage frequencies; (*a*) stage of the journey that respondents check for service information, (*b*) Social media users by age, (c) Purpose of traveling by TW-Metro; (*d*) Frequency of TW-Metro use.

age group to use social media rather than the distribution of the TW-Metro ridership. For instance, the shares provided by the NEXUS PTE in terms of journey purpose are 40% for work, 20% for education, and 40% leisure. According to the ridership statistics, adult share (85–90%) is considerably high compared with child share (10–15%) during 2011–13 (*32*).

The survey asked the main purpose for using the TW-Metro, as shown in Figure 2*c*; 60% reported they primarily traveled for work, with the other three options making up the remainder. Finally, Figure 2*d* shows the breakdown of how frequent the respondents used the metro, with almost half of the users reporting 5 + times per week.

Analysis

This section uses principal component analysis (PCA) followed by cross-tabulation analysis to identify interrelationships that emerge from the survey data. Each of these steps is presented in detail below. The goal is to gain an in-depth understanding of TW-Metro users and their use of TW-Metro social media pages.

Principal Component Analysis

To explore the relationships among the variables and to understand the data structure better, a dimension reduction process was conducted. The dataset collected for the study covers a variety of information that is related to the context of social media, including the frequency and the purpose of use, in addition to socio-demographic variables. A PCA was selected to reduce the number of observed variables to a smaller number of principal components which account for most of the variance of the observed variables (*33*). In other words, the main purpose of this form of analysis is to determine if a large

Table 1. Output from the Principal Component Analysis—Respondents' Perceptions and Usage Habits of TW-Metro Social Media Services

Statement	C	M	SD
Factor 1 Daily information update (α : 0.930)			
TW-Metro Facebook and Twitter provides me with a satisfactory number of updates in a day	0.18	3.40	1.58
TW-Metro Facebook and Twitter posts are written in a friendly yet professional manner	0.50	3.60	1.62
I understand the terms of TW-Metro used to define any disruption (minor/major)	0.16	3.41	1.66
TW-Metro Facebook and Twitter provides me with an instant reason for a disruption	0.18	3.06	1.63
Factor 2 Customer service indicator (α : 0.986)			
Time taken to receive a reply	0.20	0.61	1.39
Relevance of the answer to your question	0.44	0.68	1.52
Manner/Friendliness of the written response	0.37	0.75	1.63
Factor 3 Disruption information updates (α : 0.936)			
I am given enough notice prior to a closure through TW-Metro Facebook and Twitter	0.56	3.73	1.39
The information provided by TW-Metro Facebook and Twitter is satisfactory for me to plan an alternative journey	0.09	3.60	1.37
Enough reminders about the closure are provided through TW-Metro Facebook and Twitter updates	0.35	3.73	1.32
KMO: 0.84			

Note: KMO = Kaiser–Meyer–Olkin; α = Cronbach's alpha; C = coefficient; M = mean; SD = standard deviation.

group of variables can be adequately explained by a smaller number of unobserved constructs, referred to as factors.

The PCA specification follows some specified types of rotation methods, with the constructs being identified based on those that exceed an eigenvalue of one. Meanwhile, the initial output is subjected to rotation which alters the axis position with the aim of making the output clearer, more pronounced, and thus, further revealing the embedded structure. Tabachnick et al. suggested that the commonly used criterion of rotation selection for factor analysis is that if any of the absolute value of factor correlation matrix is greater than 0.32, "Promax" rotation should be selected (*34*). In contrast, if the absolute value is smaller than 0.32, "Varimax" rotation should be selected. Meanwhile, "Promax" rotation is recommended first in order to identify the absolute value of factor correlation matrix. Factor correlation matrix shows that the absolute value of the correlation between two factors considered in the analysis is 0.61, and since that is greater than 0.32, the "Promax" method was selected for the factor analysis.

The Kaiser–Meyer–Olkin (KMO) measure of sampling adequacy alongside Bartlett's test of sphericity is calculated to consider whether the variables are suitable for structure detection (*35, 36*). Sarstedt and Mooi [*37*]). guided that KMO less than 0.5 is unacceptable of sampling adequacy and over 0.80 is meritorious for adequacy of the correlations. To evaluate the reliability of the factors identified in the PCA, Cronbach's alpha is calculated to consider the internal consistency of the grouped statements (*38*). Table 1 presents the output from the PCA of those variables measuring respondents' perceptions about TW-Metro social media services.

According to the results, the KMO was 0.84, which is in the very good range, showing adequacy of the correlations. Exploring the output of the PCA, three factors have been identified to represent metro user perceptions toward the social media services: "daily information update," "customer service indicator," and "disruption information update." Factor 1 (daily information update) includes four statements based on the number of updates as well as the usefulness, clarity, and relevance-related aspects. Three main statements together explain Factor 2 (customer service indicator); they include Facebook and Twitter users' satisfaction in terms of time taken to respond, how relevant the response was, and the way that the response was written. Factor 3 (disruption information update) is composed of three statements that make the customers aware of the situation in case of disruption, especially in terms of advance notice for them to plan for an alternative to the journey that has already been disrupted. Peterson (*39*) indicated that the acceptable alpha scores range from 0.5 for preliminary analysis to 0.9 for applied research. In this paper, Cronbach's alpha has been calculated for each factor

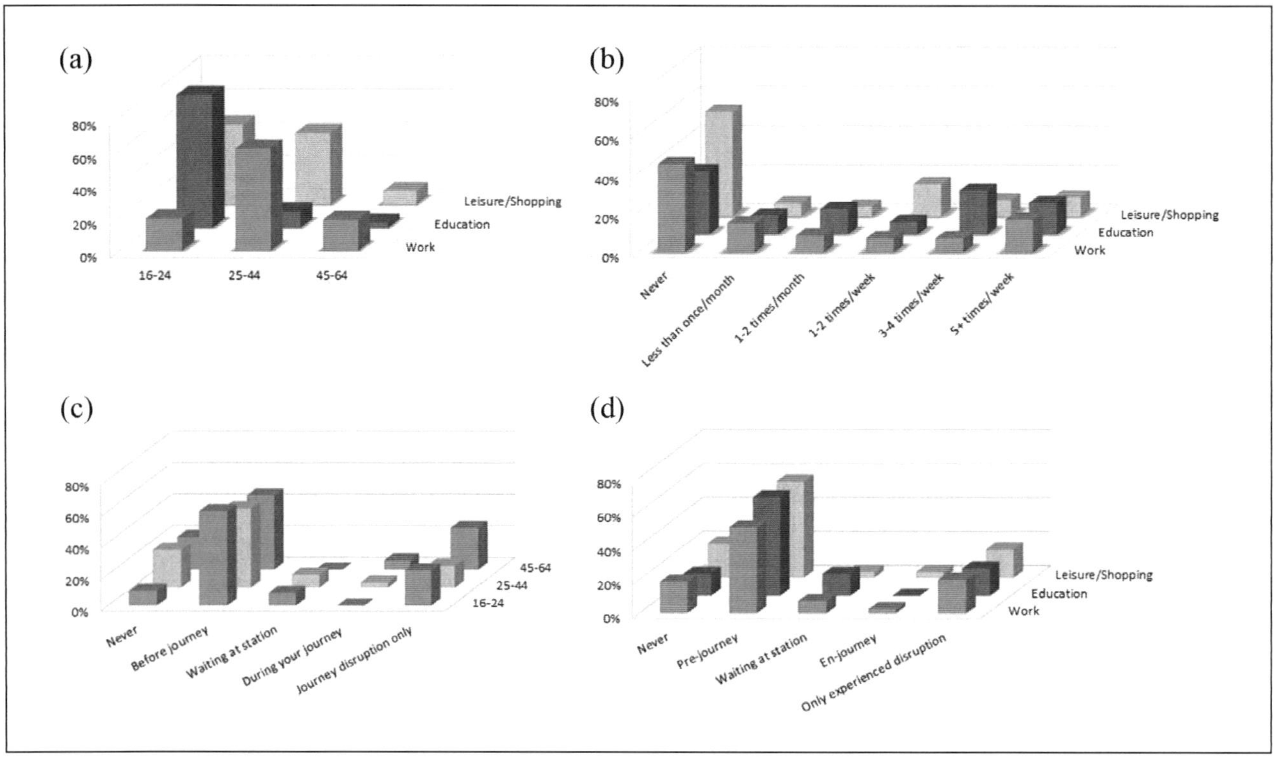

Figure 3. Segmented percentage distributions: (*a*) Respondents' age and main travel purpose; (*b*) Respondents' frequency of checking the TW-Metro social media pages and main travel purpose; (*c*) Respondents' age and the stage of the journey where they are most likely to check the TW-Metro social media pages; and (*d*) Respondents' main travel purpose and the stage of the journey they check TW-Metro Twitter/Facebook feed.

identified in the analysis. The Cronbach's alpha values are 0.93, 0.99, and 0.94, all falling above the threshold, indicating a high level of reliability.

Cross-tabulation Analysis

Cross-tabulation is very useful when examining relationships within a dataset that might not be readily apparent when analyzing total survey responses. Another advantage is that it provides a way of analyzing and comparing the outcomes for one or more variables with the outcome of another variable(s). The purpose of using the cross-tabulation analysis in this study was to investigate the relationship between travelers' age, travel purpose, and the level of usage with social media travel feeds. The outcome of the cross-tabulation analysis was aimed to test the following hypotheses using a Chi-square test (recall that the survey asked which single stage is the most common stage of the journey they check the pages for service information):

H1: there is a significant relationship between respondents' age and their main travel purpose
H2: there is a significant relationship between respondents' main travel purpose and the frequency of checking the TW-Metro social media pages
H3: there is a significant relationship between respondents' age and the most common stage of the journey when they check the TW-Metro social media pages
H4: there is a significant relationship between respondents' main travel purpose and the most common stage of the journey when they check the TW-Metro social media pages

Figure 3*a* shows that respondents belonging to the age groups 16–24 and 25–44 mainly used TW-Metro for education and work purposes, respectively. Chi-square test was used to test hypothesis H1 with a resulting *p*-value of 0.00, indicating that there is a significant relationship between respondents' age and their travel purpose. Upon testing H2, correlation between respondents' main travel purpose and their frequency of checking the TW-Metro social media pages, no significant relationship was found based on the *p*-value (0.06). The failure to detect a significant relationship may be because of the way that the age groups were proposed in the questionnaire design. For example, the structure of the age groups will include a mix of students/professionals or professionals/retirees. Figure 3*b* shows that those who travel for leisure/shopping check the TW-Metro Facebook/Twitter feed 1–2 times per week; this makes sense given the usual frequency of shopping activity is 1–2 times per week in general. A recent survey in the UK reported that 78% of the population belong to this category (*32*). Whereas, respondents who travel for educational purposes and for work purposes check the TW-Metro Twitter/Facebook feed 3–4 times per week and 5 + times per week, respectively.

There is a significant relationship that exists between respondents' age and the stage of the journey that they check TW-Metro Twitter/Facebook feed, as the resulting *p*-value of the Chi-square test is 0.03. Figure 3*c* clearly shows that respondents check the TW-Metro Twitter/Facebook feeds before starting the journey across all age groups; Figure 3*d* shows a similar trend across all three travel purposes. This indicates that the Facebook and Twitter feeds have the potential to change TW-Metro users' travel plans before their journey, such as changing their travel time, route, or travel mode. However, respondents who traveled mainly for educational purposes seemed to be checking Twitter/Facebook feeds while they are waiting at the station during transit. As the *p*-value of the Chi-square test is greater than 0.05, there is no particular correlation detected, and therefore H4 is rejected. However, additional granularity when forming the age groups as well as securing some vital information regarding the status of the users, e.g., using "students", "professionals", and "retired persons", in the questionnaire design, might have proven more useful when drawing insights for further discussions on H3 and H4.

Discussion and Conclusions

This study investigated the uses of social media for travel planning on a transit system with particular attention to travel disruptions and delays. Social media and other online services can provide general system information at low cost compared with other media. More importantly, rather than having a one-size-fits all traveler information system, these online services have the potential to provide personalized information tailored to the individual or route they are traveling. Key to this personalization is understanding the audience and their needs. The study was conducted on the TW-Metro system in Tyne and Wear, UK, but most of the results transcend the specific location or transit agency. The core of this study was a survey to explore how, and at what level, TW-Metro users utilize real-time travel information from the social media sites maintained by TW-Metro.

The original survey recruited participants via the TW-Metro social media sites, so the results cannot be used to compare different sources of information, for example, via the TW-Metro web page or sites that integrate information across different transit operators (e.g., TW-Metro does not operate buses, but there are a number of travel planning sites that integrate TW-Metro and bus trips for the purpose of travel planning in the Tyne and Wear region). Although 93% of the respondents were aware of the TW-Metro social media pages, only 40% visit them at least three times per week and roughly 14% never visit them at all. The numbers increase somewhat after accounting for frequency of travel on TW-Metro, but never climb above 57%. As most of the questions in the survey assumed regular use of social media, the 14% that never visit the TW-Metro social media sites were excluded from further processing. Presumably some of these users were lost to other information services, for example, TW-Metro's own web page. Across the demographic groups the 16–24 year olds and those commuting for education showed the greatest engagement with the TW-Metro social media sites. This result is not surprising as these groups also represent a larger share of social media consumers.

The vast majority of respondents primarily checked the social media sites to gather daily updates before starting their journeys, and this finding was true across all of the demographic groups. Thus, the TW-Metro social media sites already have the potential influence users' travel plans before their journey, such as changing their route, travel mode, and/or departure time. Meanwhile, less than 30% of the respondents sought more information related to disruptions and delays beyond the pre-trip forecasts. This outcome indicates that an active engagement with social media is still missing from the viewpoint of customers. At least part of this social media drop-off after the trip-planning stage reflects the fact that the TW-Metro system has up-to-date active message signs in the stations and audible announcements throughout the system to inform travelers about any delays or disruptions. For transit systems that do not have such standalone communication infrastructure (e.g., most bus systems), there is the potential for even greater traveler engagement, with the ultimate goal of improving user satisfaction.

As there are so many users that visit the social media pages for trip planning, there is an opportunity to reach these individuals and provide a much more dynamic engagement. Possible examples that would add this value include allowing users to create profiles to list their preferred routes, or simply providing information about transit vehicles approaching the user's location. Of course, if these backend tools are developed the interface should not be limited to social media sites that may require login; to have the greatest impact, any customized traveler information should also be accessible on the web or through a smart phone app.

Acknowledgments

The research presented in this paper was made possible by a postgraduate studentship funded by NEXUS PTE. The authors would like to acknowledge the support received from DB Regio Tyne and Wear Limited, operator of the Tyne and Wear Metro at the time of the survey, for the publication of the online survey. Also, the authors would like to thank Mr Huw Lewis, Customer Services Director at NEXUS PTE, for sharing very valuable sources of information to be included in this paper. The authors are grateful to the four anonymous referees for their detailed constructive comments and suggestions that significantly improved the quality of the paper.

Author Contributions

The authors confirm contribution to the paper as follows: study conception and design: Joanne Douglas, Dilum Dissanayake; data collection: Joanne Douglass, Dilum Dissanayake; analysis and interpretation of results: Joanne Douglass, Dilum Dissanayake, Benjamin Coifman, Weijia Chen, Fazilatulaili Ali; draft manuscript preparation: Joanne Douglass, Dilum Dissanayake, Benjamin Coifman, Weijia Chen, Fazilatulaili Ali. All authors reviewed the results and approved the final version of the manuscript.

References

1. Ofcom. *Social Networking Report*. [Online], 2008. http://stakeholders.ofcom.org.uk/binaries/research/media-literacy/report1.pdf.
2. Smith, K. *Taming the Social Media Beast*, 2013. http://www.railjournal.com/index.php/blogs/kevin-smith/taming-the-social-media-beast.html
3. CILT. *Getting the Info on Passenger Information Systems*, 2006. www.ciltuk.org.uk.
4. Beck, M. J., and J. M. Rose. The Best of Times and the Worst of Times: A New Best–worst Measure of Attitudes Toward Public Transport Experiences. *Transportation Research Part A: Policy and Practice*, Vol. 86, 2016, pp. 108–123.
5. Tyne and Wear Local Transport Plan. *Keep Tyne and Wear Moving LTP3: The Third Local Transport Plan for Tyne and Wear Strategy 2011–2021*, 2011. http://www.tyneandwearltp.gov.uk/wp-content/uploads/2011/04/TW-LTP3-Strategy-Mar-2011-for-upload.pdf.
6. Grotenhuis, J.-W., B. W. Wiegmans, and P. Rietveld. The Desired Quality of Integrated Multimodal Travel Information in Public Transport: Customer Needs for Time and Effort Savings. *Transport Policy*, Vol. 14, No. 1, 2007, pp. 27–38.
7. Community of Metros. *Community of Metros Case Study 2010: Information During Disruptions Final Report*. Imperial College London, London, England, UK, 2011.
8. Beirão, G., and J. A. Sarsfield Cabral. Understanding Attitudes Towards Public Transport and Private Car: A Qualitative Study. *Transport Policy*, Vol. 14, No. 6, 2007, pp. 478–489.
9. Bluegrass. DB Regio Tyne and Wear Communication Service Disruption Focus Group Findings, 2012, unpublished report.
10. Transport Focus. The Future of Transport, 2012. https://www.transportfocus.org.uk/research-publications/publications/the-future-of-transport-overview/. Accessed on July 20, 2016.
11. Nexus. *Metro Strategy 2030 Background Information*, 2013. http://www.nexus.org.uk/sites/default/files/Metro%20Strategy%20Background%20document.pdf
12. Kuzma, J. Asian Government Usage of Web 2.0 Social Media. *European Journal ePractice*, Vol. 9, No. 3, 2010, pp. 1–13.
13. Landsbergen, D. Government as Part of the Revolution: Using Social Media to Achieve Public Goals. *Electronics Journal of e-Government*, Vol. 8, No. 2, 2010, pp. 135–147.
14. Cairns, S., S. Aitkins, and P. G. Goodwin. Disappearing Traffic? The Story So Far. *Proc., Institution of Civil Engineers; Municipal Engineer*, Vol. 151, No. 1, 2002, pp. 13–22.
15. Van Excel, N., and P. Reitveld. Public Transport Strikes and Travel Behaviour. *Transport Policy*, Vol. 8, 2001, pp. 237–246.
16. Caulfield, B., M. O'Mahony, and S. Farrell. An Examination of the Public Transport Information. *IEEE Conference on Intelligent Transportation Systems*, Vienna, Austria, 2005.
17. Ambrosino, G., J. D. Nelson, and M. Romanazzo. *Rome: ENEA*, 2004. http://www.enea.it/en/publications/volume-pdf/V2004-Demand_Transport_Services.pdf.
18. Iryo, T., K. Yamabe, and Y. Asakura. Dynamics of Information Generation and Transmissions Through a Social Network in Non-Recurrent Transport Behaviour. *Transportation Research Part C: Emerging Technologies*, Vol. 20, No. 1, 2012, pp. 236–251.
19. Camacho, T. D., M. Foth, and A. Rakotonirainy. Pervasive Technology and Public Transport: Opportunities Beyond Telematics. *IEEE Pervasive Computing*, Vol. 12, No. 1, 2013. http://ieeexplore.ieee.org/stamp/stamp.jsp?tp = &arnumber = 6336696.
20. Nunes, A. A., T. Galvao, J. Falcao, e Cunha, and J. V. Pitt. Using Social Networks for Exchanging Valuable Real Time Public Transport Information Among Travelers. *IEEE Conference on Commerce and Enterprise Computing*, Vol. 2011, 2011, pp. 365–370.
21. Ofcom. *Adult Media Use and Attitudes*, 2016. http://stakeholders.ofcom.org.uk/binaries/research/media-literacy/adults-literacy-2016/2016-Adults-media-use-and-attitudes.pdf.
22. Boyd, D. M., and N. B. Ellison. Social Network Sites: Definition, History, and Scholarship. *Journal of Computer-Mediated Communication*, Vol. 13, No. 1, 2007, pp. 210–230.
23. Facebook. *News Room*, 2014. http://newsroom.fb.com/company-info/. Accessed on July 27, 2017.
24. Austin, J. *Use of Social Networking to Promote Public Transport and Sustainable Travel*. [Online], 2013. http://www.analytics.co.uk/resources/Use+of+Social+Media+to+promote+PT+$26+Sustainable+Travel.pdf. Accessed June 21, 2013.
25. Social, A. *The Latest UK Social Media Statistics for 2016*, 2016. http://avocadosocial.com/1095-2/. Accessed on July 27, 2017.
26. Clifton, P. Watch the Birdie. *Rail Professional Magazine*, XEIAD Publishing, 2011.

27. Baird, C. H., and G. Parasnis. From Social Media to Social Customer Relationship Management. *Strategy & Leadership*, Vol. 39, No. 5, 2011, pp. 30–37.
28. Neuxs. *How Metro was Built*, 2017. https://www.nexus.org.uk/history/how-metro-was-built. Accessed on July 20, 2017.
29. Nexus. *Tyne and Wear Metro Routes*, 2017. http://www.thetrams.co.uk/tyneandwear/routes.php. Accessed on July 20, 2017.
30. Smith, A. and D. Mais. Light Rail and Tram Statistics: England 2016/17, 2017. London: Department for Transport, 2017. https://assets.publishing.service.gov.uk/government/uploads/system/uploads/attachment_data/file/623366/light-rail-tram-ending-march-2017.pdf
31. Quintly. *Response Rates by Facebook Pages Kept Growing in 2014*, 2014. https://www.quintly.com/blog/2015/01/response-rates-facebook-pages-2014/
32. Nexus. *Nexus Strategic Intelligence Report 2012-13*, 2013. https://www.nexus.org.uk/sites/default/files/Strategic%20Intelligence%20Report%202012-13%20(Alex%20Finkel).pdf
33. Suhr, D. D. *Principal Component Analysis (PCA) Vs Exploratory Factor Analysis (EFA)*, 2016. www2.sas.com/proceedings/sugi30/203-30.pdf
34. Tabachnick, B. G., L. S. Fidell, and S. J. Osterlind. *Using multivariate Statistics*, 2001.
35. Bartlett, M. S. Tests of Significance in Factor Analysis. *British Journal of Statistical Psychology*, Vol. 3, No. 2, 1950, pp. 77–85.
36. Hotelling, H. Analysis of a Complex of Statistical Variables into Principal Components. *Journal of Education Psychology*, Vol. 24, No. 6, 1933, p. 417.
37. Sarstedt, M., and E. Mooi. *A Concise Guide to Market Research: The Process, Data, and Methods Using IBM SPSS Statistics*. Springer, Berlin, Germany, 2014.
38. Cronbach, J. L. Coefficient Alpha and the Internal Structure of Tests. *Psychometrika*, Vol. 16, No. 3, 1951, pp. 297–334.
39. Peterson, R. A. A Meta-analysis of Cronbach's Coefficient Alpha. *Journal of Consumer Research*, 1994, pp. 381–391.

The Standing Committee on Effects of Information and Communication Technologies (ICT) on Travel Choices (ADB20) peer-reviewed this paper (18-03942).

TRR
JOURNAL OF THE TRANSPORTATION RESEARCH BOARD

Article

Transportation Research Record
2018, Vol. 2672(50) 56–65
© National Academy of Sciences:
Transportation Research Board 2018
Article reuse guidelines:
sagepub.com/journals-permissions
DOI: 10.1177/0361198118792753
journals.sagepub.com/home/trr
SAGE

Ex-Ante and Ex-Post Evaluation of a New Transit Information App: Modeling Use Intentions and Actual Use

Guillermo Velazquez[1], Sigal Kaplan[2], and Andres Monzon[3]

Abstract

This study investigates the behavioral drivers underlying the adoption of a multimodal travel information mobile app. The hypothesized framework is validated empirically through the case-study of Madrid. Madrid's Public Transport real-time information app ("Mi Transporte") allows users to obtain customized and automated information. A three-wave survey containing questions aligned with the Theory of Planned Behavior was conducted in 2015 and 2016 with a representative sample of transit users. Data analysis includes a factor analysis and a structural equation model to validate the hypotheses. The model assumes that the intention to use the app can be explained as a function of attitudinal factors and respondent characteristics. Results show that the app adoption is correlated with the intention of the users to adopt it and with their willingness-to-pay; the users' intentions can be explained by various factors like user's expectations on the app, affinity for technology (technophilia) and the previous use of other transport apps. The roles of search functionalities, side-mode information, time saving skills and the importance of the Level of Service (LOS) are also analyzed in the model. Relations between user characteristics and latent variables are subsequently explained as well as the ex-post satisfaction and change in travel patterns to measure the impact on the transport behavior of the app users. The study provides a better understanding of app adoption based on traveler characteristics, the attributes of the app and the perception of its capabilities.

As large centers of human and economic activity, modern metropolitan areas and mega-cities are characterized by population density and infrastructure design that define the urban efficiency level based on mobility and accessibility, considering the city area and the transport infrastructure by mode. While more than half a century ago, people agreed that cities had the capacity to expand, it is becoming clearer that this capacity is becoming exhausted. Thus, the only viable means for generating sustainable and competitive urban environments is to increase the mobility efficiency provided by the existing supply while maintaining regional growth. To enhance the efficiency of modern cities on the basis of the existing infrastructure, travelers must have the ability to combine different means of transport, adapting their choices to their circumstances as well as to the city context. Multimodal transport systems need to operate as integrated, seamless networks. In this regard, multimodal trips have increased steeply in recent years, being one of the objectives of the European Commission for the future mobility of the EU's urban areas and evidence of generations Y and Z becoming multimodal in several countries in northern Europe. Rethinking travel efficiency in the digital age, new ICT solutions have been introduced to enhance integration between modes at the information level. There is a proven positive impact of improved transit information on urban economies, mobility and environment as it results in more efficient journeys for individual travelers, and improved satisfaction with the service (*1*). Moreover, Ibraeva and Sousa (*2*) pointed out that the lack and/or complexity of information about routes and timetables serve as powerful barriers for potential transit use. Thus, providing easily accessible and trustworthy information sources is essential for enhancing the use of low-carbon modes of transit and increasing the efficiency of emission and fuel consumption per passenger trip. While information technology is relatively cost-efficient and its use is spreading

[1]Transport Research Centre (TRANSyT), Universidad Politecnica de Madrid, Madrid, Spain
[2]Department of Geography, Hebrew University of Jerusalem, Mount Scopus, Jerusalem, Israel
[3]Universidad Politecnica de Madrid, Madrid, Spain

Corresponding Author:
Address correspondence to Guillermo Velazquez: g.velazquez@upm.es

rapidly among transport users, the development of ICT solutions is rarely financially supported by transport authorities and operators, most likely because the effective reach of mobile technologies and their impact on travel patterns are yet unclear. Understanding the relation between real time information systems and transit use could support transit operators in their decision to invest in advanced information systems

Objective measurable indicators support a positive relationship between information provision and transit ridership as the outcome. Interestingly, information still plays a minor role in modal choice when compared with journey planning and execution for a chosen mode (*3*). Today individuals still have to make their travel decisions in uncertain circumstances with respect to travel time; they are not able to predict the exact travel time or arrival time before starting their trips given a departure time (*4*), this uncertainty increases in the case of multimodal chains. Yet studies show the potential of travel information systems to change travel patterns as a result of travel-time savings (*5*), reducing travel-time variability (*6*) and increasing user satisfaction (*7*). While little research addressed the link between information and transit use, studies show a positive relationship between transit information and ridership (*8–10*) the same as for cycling (*11*). Therefore, the use of adequate channels to reach both active and potential users with information could be an important policy instrument. The utility of travel information provision becomes tangible in the users' Willingness-To-Pay (WTP) for travel information, as demonstrated for example in Greece, the UK, and China (*12*). Users are willing to pay for information that provides added value compared with existing information services (*13*) and that the most valued feature is real-time provision (*14*).

This research provides new insights into the willingness to adopt smartphone-based travel information apps. Until recently, pre-trip information could either be gathered at home or en route at designated locations (transit stops, vehicles etc.). But now that travelers carry their information source as they travel and real-time information is continuously updated during the trip. Precise spatial-temporal information at any time, any place about every transport mode could be a game changer because it allows travelers to plan, change, re-schedule, and chain trips spontaneously and 'on the go'. This advantage may greatly facilitate multimodal and sustainable travel modes. The general increase in market penetration rates of mobile devices from developing to the most advanced economies, equipped with user interfaces, suggests that smartphones could serve as a global opportunity in user–operator communication. Transport authorities now have the opportunity to achieve better integration of transport modes to facilitate multimodal transport choices in an effective and cost-efficient way by a simple-to-use push-button app. Mobile applications appear as a new solution to tackle these information provision problems, presenting travelers with a personalized set of modal options that can include soft aspects like comfort or convenience, which can be determinant for a modal change (*3*). Khattak et al. (*15*) found that individuals are more willing to pay for travel information via the telephone when the information is customized, their trips are longer, and the trip is related with work. The actual adoption of these mobile travel information apps is crucial for their deployment and improvement. In situations involving new technologies, intentions to use form a rigorous measure of future use behavior.

This study uses a three-wave panel survey to investigating the actual use of a new app throughout the two-year research period. The Extended Technology Acceptance Model (ETAM) served as the behavioral framework for data analysis, further elaborated from previous studies by linking technology orientation, WTP, and user characteristics and perceptions. Doing so, this study extends existing knowledge both on app adoption (*16*) and usability (*17*). Up to now, little attention has been paid to the role of behavioral constructs in the association between expectations, intention to use the new information system, the WTP, and the actual adoption of mobile public transportation apps. Regarding the scope, the research analyzes the latent constructs and variables related to user adoption of a new multimodal real-time information app for public transport in Madrid. In the following section the applied methods are described: experimentation plan, proposed behavioral framework, and mathematical model. Then the case study of "Mi Transporte" app is introduced. Finally, the results and conclusions of the analysis are presented. The study adds to previous studies by validating the hypothesized behavioral framework and the structural relations among variables for the case study of Madrid (Spain).

Methodology, Implementation, and Analysis

Madrid is characterized by a highly integrated and well-structured multi-modal transit system, where over one third of total daily trips are carried out on public transport and with a highly technophile user base. However, until recently there was no single official channel integrating real time information from the various transportation modes available to citizens (urban bus, metro, interurban bus, railway, walking). In May 2016, Madrid implemented its first official Public Transport real-time information app ("Mi Transporte" http://www.crtm.es/atencion-al-cliente/area-de-descargas/apps/app-tiempo-real.aspx). Many cities have already done so, e.g., the apps developed by Transport for London, the Paris Region, or Singapore on the public side, or the opening of data to the private sector that has developed apps in cities like San Francisco, Boston, or London. "Mi Transporte" provides general information, such as the nearest stops, line routes, arrival times, and trip disruptions, as well as the possibility for the user to obtain customized or automated information.

Hypothesized Behavioral Framework

The methodological procedure for the study starts with the development of hypothesis on the behavioral framework underlying the adoption of a transit mobile app. A survey is then designed, taking into account the needs of the model, to gather the relevant information on a representative sample of potential app users. The hypothesized behavioral model structure has been statistically tested and validated through a mathematical model. For our study, Structural Equation Models (SEM) were applied. SEM are a well-established and widely-applied method in behavioral, social and natural sciences, they allow us not only to parameterize correlations among variables but to fully understand them and thus to produce improved policy recommendations.

We hypothesize that the use of the app is related to the WTP for the app and the interest in trying it. This interest is motivated by personal technology affinity (technophilia), time saving skills, environmental responsibility, perceived importance of the level of service (LOS) and satisfaction with interchanges (the study was performed in one main interchange station), information search needs on transit, bicycle and car, and expectations regarding the benefits of the app. All of these elements were addressed in the questionnaire through specific sets of items.

The proposed behavioral framework is based on the Extended Technology Acceptance Model (ETAM), adapted to the context of transit use and mobile app use. In the ETAM, the demonstrability of the results and the output quality explain the perceived technology usefulness of the app which is expressed in terms of time and monetary savings. Subjective norms have not been considered as the usage of the app is not visible for other users. User experience or familiarity have been considered through the previous usage of mobility apps and the familiarity with technological devices. The construct of job relevance in the ETAM is a contextual variable and therefore irrelevant in the current app use context. Therefore, the proposed model framework enables us to explore the following hypotheses:

H1: app use is related to the willingness-to-pay for the app and the intention to use it.

H2: user's expectations, technophilia, and previous use of transport apps explain the intention to adopt the app.

H3: the search functionalities of the app, the LOS importance, information about side-modes and time saving skills explain the user's expectations.

H4: gender, age and smartphone use explain user's technophilia.

H5: trip type (frequency, duration, reason of trip) and smartphone use are related to time saving skills.

Mathematical Structural Equations Model (SEM)

SEM presents a general statistical approach for validating a multivariate correlation structure across latent and observed variables in a series of equations. SEM is a generalization of regression that has an advantage in accommodating (i) a correlation structure across multiple dependent variables, and (ii) conceptual (latent) constructs that cannot be perfectly measured by a single observed indicator, but require multiple indicators. SEM is commonly viewed as a combination of confirmatory factor analysis (CFA), representing the relationship between the latent constructs and the set of observed indicators, and path analysis representing the dependency structure across the latent constructs (*18*). A growing number of recent studies applied SEM to explore mode choice and the mode switching intentions of adults (*19–21*). The methodology is thoroughly presented by Pugesek et al. (*18*), and its application in travel behavior research has been reviewed by Golob (*22*). In this study structural equations are used to explain the intention to use the app as a function of attitudinal factors and observed respondent characteristics.

The commercial software M-Plus served for the model estimation. The parameters were estimated simultaneously by using Maximum Likelihood with Huber-White covariance adjustment (*23*). Standard errors were calculated by adopting the White's sandwich-based method that produces robust statistics in the presence of non-normality of the indicators and the categorical variables (*24–26*).

Survey Design and Administration

A survey was conducted during 2015 and 2016 to investigate the relevant variables affecting the attractiveness and acceptance of the app. To obtain a representative sample of transit users (potential app users) we applied a mix survey methodology based on a personal intercept short interview combined with a web-based survey, a technique tested by Hernandez, et al. (*27*). More than 2,000 public transport users of different ages, using different transportation modes and at different days of the week and hours of the day were intercepted and asked some questions and then asked to fill in the online survey. Those completing the web questionnaire entered a prize draw as an incentive. They were intercepted in Avenida de América, one of the main interchange stations in the city that combines four subway lines (one circular line and three others), eleven urban bus lines, and twelve inter-urban bus lines, as well as park and ride facilities, handling approximately 170,000 travelers per day.

The questionnaire was focused on five aspects: (i) use of Public Transport, (ii) use of technological devices, (iii) Perceptions on PT and on PT apps, (iv) willingness-to-pay for an app with certain characteristics (those to be introduced) and characteristic rating (WTP), and (v) user socioeconomic characteristics. It was a panel survey conducted in three waves. The survey consisted mainly of questions associated with the Theory of Planned Behavior that states that attitudes toward behavior, subjective norms, and perceived behavioral control, together shape an individual's behavioral intentions and behaviors (*28*). Further elements of the online survey and the survey methodology have been based on the practice of social research (*29*). The questions were rated on a five-point scale from "strongly disagree" to "strongly agree." The questionnaire design included both positively and negatively phrased items. Reverse-scoring was employed as the standard procedure for data analysis to combine the negatively and positively phrased items.

The survey included items intended to observe the users' characteristics in travel behavior, technological skills, and predisposition to adopt both new technological solutions in general and mobile apps in particular. Completed by a set of questions on the users' preferences for a transportation mobile app, as well as on their willingness to pay for such an application. In total 386 people answered to the *ex-ante* survey and 153 remained through the whole panel study.

Figure 1 illustrates the three waves of the survey. Wave 1 (*ex-ante*) selected the users distributed by the different sample categories, before the app "Mi Transporte" was launched. Wave 2 was carried out one month after the app was put in use, by which time users had started to get used to it and to realize its utility. Finally, Wave 3 sought to capture changes in travel behavior and to assess the value added by the app.

Case Study: Madrid

Madrid is a city of some 3.5 million inhabitants with its metropolitan area reaching 6 million. It covers a land-developed area of 1,037 km^2. Three circular motorways surround the city (M-30, M-40 and M-50) giving access along seven radial motorways. It has five fully operational (out of seven projected) multimodal interchange stations that connect the main accesses to the city with the public transportation network. According to the Household Mobility Survey of the Madrid Region (conducted by the Madrid Public Transport Authority (CRTM) in 2014) about 5 million displacements on public transport occur on a working day in the whole metropolitan area. This fact gives an idea of the importance of public transport management in the city and the metropolitan area. The CRTM manages 700 regional bus lines, 400 urban bus lines, 12 railway lines, 16 metro lines, as well as six light rail lines. Within Madrid, two-thirds of daily trips are carried out in sustainable modes, half of them on public transport and the other half on foot.

Madrid City and the CRTM are improving sustainable mobility by further enhancing public transportation and multimodality. Madrid City introduced in 2015 a new public bike-sharing system (BiciMAD) in the central area of the city which has continued to grow the area it covers and seen the number of users increase. In their same vein, CRTM is also working on new solutions that exploit their operation data to improve the information services provided to the users and to reduce user uncertainty associated with route planning and disruptions in service caused by unexpected incidents.

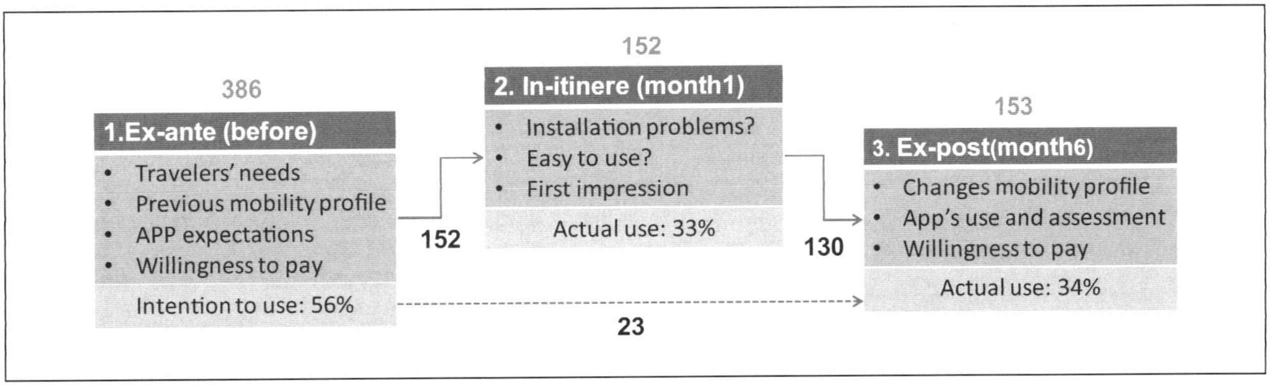

Figure 1. Panel Survey Evolution through the three waves.

Figure 2. "Mi Transporte" outlook. Configuration and map screens.

Regarding information provision, the main public transport operators have launched in recent years information channels. Madrid city-bus (EMT) and Metro companies launched their real time information apps, both available on the web from 2014 and on mobile devices from 2016, but they are not integrated. Also in 2015–2016, EMT installed panels providing real time information about approaching services at each bus stop. Metro did the same at station entrances and platforms. At the multimodal level, both Madrid City and CRTM launched their Open Data portals containing real time as well as historic transit data for developers to integrate such information in diverse solutions for PT users.

The app subject to this analysis has multimodal dimensions, combining static and real time information about the transport supply offered by 40 different public transport operators in the Madrid Metropolitan Area and City. It improves the communication of the information on the transport network to the final users, helping them to find the closest stop, providing the next train/bus arriving at that stop, informing them about any disruption to their trip, etc. See Figure 2.

Results and Discussion

Sample Characterization and Technophilia

The survey involved a total of 200 women and 186 men. It was stratified in three age groups: young (18–30, 24%), adults (30–45 years, 51%), and senior travelers (50–70 years, 26%). It is worth mentioning that the most limiting group to be targeted to achieve a representative sample were senior travelers, as they are less adroit in the use of technology. The travel behavior of respondents was collected through dedicated survey questions related to their most frequent trip. The results showed that, as expected, a significant proportion of trips were intermodal, with the combinations of Bus-Metro and Metro-Bus accounting for more than 63% of multimodal trips.

To ascertain the extent to which users were comfortable with the use of different technological tools, they were asked about both their current IT device usage and their proficiency level in the use of each of the more broadly extended devices: PC, Smartphone, Tablet, and GPS. The results, included in Table 1, show that technological penetration rates are high among public transport users in Madrid. Some 78% use personal computers and up to 86% declared a proficiency level; almost 84% are users of smartphones with the same rate claimed a good or very good proficiency use level.

Exploratory Factor Analysis

The data analysis included exploratory and confirmatory factor analysis followed by a structural equation model to empirically validate the framework. The objective of exploratory factor analysis is to identify a set of latent constructs underlying a set of measured variables in a situation without a priori hypotheses, while confirmatory factor analysis is to test whether the data fit a

Table 1. Technological Capabilities of the Sample by Age

	15–30 years old	30–50 years old	50–65 years old	Average
Smartphone	89.3%	82.9%	73.3%	83.7%
PC	80.3%	76.3%	76.0%	78.0%
Tablet	44.0%	49.3%	32.0%	43.8%
GPS	20.1%	27.6%	26.7%	24.4%

hypothesized measurement model. The exploratory factor analysis served to extract the latent constructs among variables. Tests of internal consistency and sample adequacy constituted the necessary preliminary conditions for conducting factor analysis and obtaining meaningful results.

The items obtained in the survey demonstrate good internal consistency (Cronbach's alpha = 0.876) and good sampling adequacy according to Kaiser-Meyer-Olkin measure (KMO = 0.881). The Spearman correlation matrix shows the existence of correlations without multi-collinearity. The result of the Bartlett's test of sphericity rejects the null hypothesis that the correlation matrix is an identity matrix ($p = 0.000$). Exploratory principal axis factor analysis with varimax rotation (Kaiser normalization) produced six or seven factors according to the Scree plot. The existence of a factor that contained information on environmental concerns was deemed interesting for the analysis and therefore seven factors were taken into account for the modeling. The interpretation of the meaning of the factors relys on their supporting questions scored by the respondents. Thus, the first factor (F1) has been labelled "Technophile". The next factor (F2) "Expectations regarding the app", the third (F3) is labelled "app search functionalities." The next three factors were labelled "Level of Service (LOS) importance" (F4), "Side-mode information" (F5), and "Time saving skills" (F6). Finally, the last factor (F7) grouped "Environmental responsibility" items. Then, those factors were confirmed by confirmatory factor analysis, then contrasted, associated to user characteristics (e.g., age, gender, trip length and type, principle mode and smartphone use) and introduced into the structural relation model. The analysis of the results in Table 2 shows the explanatory power of the seven identified latent variables and the questions of the survey that support the loading factors for each of them.

Structural Equation Model Estimation and Results

The estimation results are presented in Table 3 which presents the measurement equations of the confirmatory factor analysis that corresponds to exploratory factor analysis presented in Table 2 in the previous section. In this study, the factor structure is revealed by exploratory factor analysis and it is used as part of the model structure with confirmatory factor analysis, which enhances the structural validity of the proposed model. The CFI value, which indicates the goodness-of-fit of the confirmatory factor analysis, is 0.835, showing excellent data fit. The model including the set of measurement and structural equations fits reasonably well.

The statistical model confirms the general behavioral framework and the research hypotheses for the case study of Madrid. Table 3 shows the relationship between the latent constructs and the respondents' characteristics. Most importantly, we focus on the effect of the various latent constructs on the intention to use the real-time information app, as well as on the WTP and the actual app use. The model statistically confirms H1 with the relation of WTP for the app and the intention to use it being significant for its actual adoption. The statistical model also confirms H2. With the user's expectations, the user's technophilia, and their previous use of transport mobile apps explaining the intention to adopt the app. The statistical model confirms H3. The user's expectations are positively and significantly explained by search functionalities of the app, the LOS importance, information about side-modes and time saving skills. The correlation patterns on Table 3 do confirm H4. With Gender and age negatively and significantly correlated (male and younger are more technophile) with the technophile factor and smartphone use positively and significantly correlated with the technophile factor (F1). The model also confirms H5 with the importance of Time Saving Skills positively and significantly related to the frequency and duration of the trip, as well as to the use of smartphones, while negatively and significantly related to the most frequent trip taking place inside the City Center.

Finally, the analysis also showed that the importance of *Side-Mode* information is positively and significantly correlated to the *Environmental responsibility* factor, and that the importance of the LOS is explained by *Age* and *Gende*r (female and elder give more importance to LOS). These results and the weights among interrelated factors are explained below and are represented in Figure 3.

Table 2. Output of Rotated Factor Analysis (Exploratory)

		F1	F2	F3	F4	F5	F6	F7
I like to try new technological devices.		.766	.154	.195	.099	–.023	.119	.027
I believe that new technologies have great potential.		.664	.115	.252	.074	–.120	–.017	.053
I'm interested in new technologies.		.802	.153	.194	.044	–.024	.095	.027
I like to renew my mobile from time to time to keep it updated.		.572	.128	–.045	.050	.093	–.033	.117
Apps help me in my daily life.		.720	.238	.068	.032	.086	.016	.048
I need technological tools in my daily life.		.703	.088	.008	–.073	.076	.057	.043
Some apps are fun to use.		.674	.151	.102	.138	.011	.006	.117
I like to find new applications.		.773	.195	.076	.100	–.007	.041	.099
Please indicate your level of skills in the usage of Tablet devices.		.580	.131	.257	.026	.093	.294	–.325
Please indicate your level of skills in the usage of Smartphone devices.		.610	.087	.198	.089	.034	.202	–.264
Please indicate your level of skills in the usage of Laptop devices.		.517	.128	.297	–.019	.054	.295	–.294
Please indicate your level of skills in the usage of GPS devices.		.517	.011	.153	.062	.252	.248	–.198
Please indicate the importance of each	Time	.135	–.018	.005	.601	.078	.202	-.132
of the following criteria for your	Flexibility	.111	.103	.122	.727	–.005	–.059	.117
MOST FREQUENT TRIP.	Predictability	.147	.035	.103	.570	–.113	.120	.025
	Safety	–.004	.127	.069	.540	–.064	–.025	-.022
	Security	–.014	.061	.068	.593	–.071	–.066	-.066
	Comfort	.025	.120	.039	.674	.098	.073	.033
	Pleasure	–.074	.159	–.052	.654	.118	–.145	.160
	Capability to carry items	.071	.047	.064	.569	.067	–.020	.269
My most frequent trip takes too much time.		.090	–.064	.096	.045	–.009	.681	.017
I feel like I waste my time on the way.		.049	–.042	–.016	–.081	.019	.694	–.101
I would use public transport more often if there was real-time service information.		.136	.151	–.076	.119	–.050	.510	.361
I would use public transport more often if the service was better.		.080	.027	–.089	.116	.042	.714	.264
The impact of transport on the environment is a real problem in Madrid.		–.018	.015	.123	.014	.027	.238	.681
I feel as part of my responsibilities to reduce the emission of pollutants that potentiate the greenhouse effect.		.000	.082	.000	.123	.068	.058	.786
It's important to search for the name of the stop.		.199	.151	.517	.079	–.045	.084	.033
It's important to find the nearest stop of a chosen line.		.113	.202	.723	.073	.082	–.011	.038
It's important to find the nearest stop of a chosen mode of public transport.		.139	.106	.800	.052	.081	–.034	.061
It's important to find the nearest public transport stop.		.061	.063	.758	.080	.117	.032	–.005
It's important to find the nearest BiciMAD station.		.063	.186	–.011	.020	.790	–.013	.103
It's important to be able to search for nearby BiciMAD stations according to occupancy level.		.075	.194	–.010	.015	.789	–.063	.073
It's important to be able to find deterrent parking facilities (and information on public transport services there).		.111	.067	.116	–.009	.802	.009	.008
It's important to be able to search for parking facilities according to occupancy level.		.121	.018	.097	.049	.795	.033	–.003
If the app allows me to save travel time, I intend to use it in my daily trips.		.316	.595	.445	.113	–.036	.103	.005
If the app can increase the predictability of my daily trips, I will use it.		.270	.574	.438	.169	–.080	.039	–.052
If the app allows me to save money, I intend to use it.		.302	.569	.475	–.022	–.102	.100	–.041
If the app allows me to plan my trips better, I intend to use it.		.291	.565	.432	.036	–.039	–.034	–.031
Using the app will allow me to save some travel time.		.143	.788	.100	.119	.022	.012	.013
Thanks to the app I will get to my destination more calmly.		.145	.800	.060	.038	.128	.002	.037
Thanks to the app I will improve the regularity in my times of displacement.		.155	.783	.145	.150	.127	–.055	.009
Thanks to the app I will reduce the impact of my travel on the environment.		.012	.593	.049	.003	.269	–.025	.333
I think using the app will make it easier for me to change my transport habits.		.200	.691	–.168	.061	.165	.020	.079
I think the app will be easy to use.		.319	.538	.186	.158	.012	.064	–.055

Note: The factor loadings are presented in the table. To facilitate factor labeling, the dominant items were defined as those with an absolute value of the loading greater than 0.50.

Table 3. Structural Equations Relating Latent Factors to App Use Intention and Adoption

Latent constructs	Explanatory/mediator variable	Est.	C.R.
Side-mode information (F5)	Environmental responsibility (F7)	0.873	2.954
Expectations regarding the app (F2)	App search functionalities (F3)	1.213	7.008
	LOS importance (F4)	1.204	4.450
	Side-mode information (F5)	0.094	3.924
	Time saving skills (F6)	0.327	2.580
Technophile (F1)	Age	–0.024	–3.776
	Gender	–0.294	–1.622
	Smartphone User	1.495	5.911
LOS importance (F4)	Age	0.006	2.411
	Gender	0.201	3.001
Time-saving skills (F6)	Smartphone User	0.187	1.331
	Frequency of most frequent trip	0.073	1.816
	Most frequent trip is long	0.255	1.485
	Most frequent trip is center-center	–0.290	–2.315
Dependent Variables	Explanatory / mediator variable	Est.	C.R.
Intention to use the app	Technophile (F1)	0.847	3.371
	Expectations regarding the app (F2)	1.104	4.088
	Previous app User	0.944	2.126
Willingness To Pay (WTP)	Willing to pay to save 40% or more of trip time	0.230	2.300
Use of the app	Willingness to Pay (WTP)	0.188	1.377
	Intention to use the app	0.098	2.548

Note: Ratio between chi-square and degrees of freedom = 2.42 which are good (*30*). RMSEA stands at 0.072. CFI stands at 0.835, which are reasonable according to Loehlin et al. (*31*).

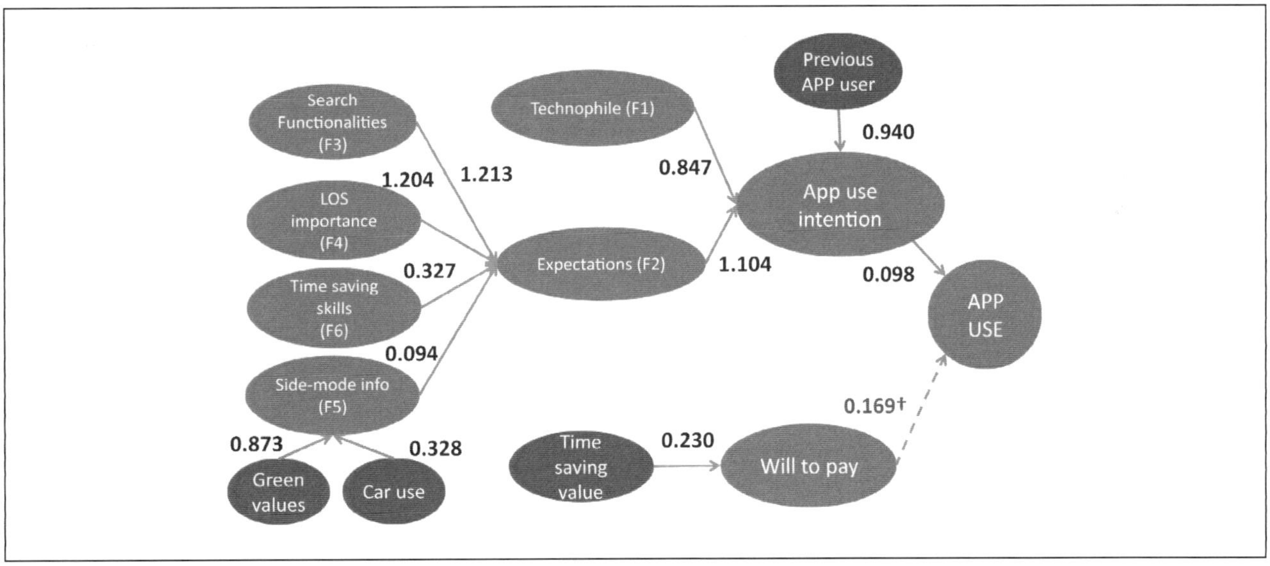

Figure 3. Structural equations model: relation among variables and latent variables.

The analysis of the SEM shows that (i) the adoption of the app is correlated with the user's intention to adopt it and more strongly with the WTP; (ii) while the intentions can be explained by various factors, the WTP is only related to the WTP to save time; (iii) a user's expectations, technophilia, and previous use of transport apps strongly explain the intention to adopt the app; (iv) the search functionalities, the LOS importance, information about park-and-ride and bicycle and time saving skills explain the user's expectations; (v) gender, age and smartphone use explain a user's technophilia, (vi) trip type, frequency, duration, and smartphone use are related to time-saving skills.

Conclusions

The "Mi Transporte" app has achieved relative success. Of the users that tested it, 35% are using it frequently for their most frequent trip and 48% of them stated that the app is now their main mobility app. For most (60%), the app has changed their waiting time or route choice. The general rating and user knowledge on the app improved along the experimentation period, with no severe reliability issues being reported. The most important declared attributes are that the app is free, multimodal and that it provides information in real time. WTP is low, and always linked to tangible time savings thanks to the use of the app.

Model results expand the body-of-knowledge from the mere analysis of attitudes to a wider scope that considers user characteristics, perceptions and behavior. As hypothesized, user expectations, technophilia, and previous use of transport apps explain the intention to adopt the app. The expectations are explained by search functionalities of the app, LOS importance for the user, availability of side-mode information and the user's time saving skills. The model also explains the actual adoption of the app, as related to WTP and the intention to use it.

These results provide researchers and practitioners with a better understanding of the use of mobile app solutions by transit users and shed light on potential elements to be considered for the implementation of transit app technologies by cities for them to achieve their maximum potential linked to personalized services offering (Mobility as a service). We have attempted to reduce potential biases, for example, by using the appropriate statistical methods, but we recognize that certain biases inherent in survey research, such as self-selection and policy response bias, may remain.

Future research lines include analyzing the effect of transit use frequency on app adoption and WTP or studying the underuse of features like automation of reiterative searches or personalized alarms revealed by the *ex-post* analysis in comparison with the declared interest in the *ex-ante* survey.

Acknowledgments

The authors gratefully acknowledge that the data employed for the study are part of the Madrid Pilot Case coordinated by Madrid Public Transport Authority (CRTM), for the project OPTICITIES (www.opticities.com), funded by the European Commission under the 7th Framework R&D Programme.

Author Contributions

The authors confirm contribution to the paper as follows: study conception and design: Guillermo Velázquez, Sigal Kaplan, Andrés Monzón; data collection: Guillermo Velázquez, Andrés Monzón; analysis and interpretation of results: Guillermo Velázquez, Sigal Kaplan, Andrés Monzón; draft manuscript preparation: Guillermo Velázquez. All authors reviewed the results and approved the final version of the manuscript.

References

1. Skelley, T., A. Namoun, and N. Mehandjiev. The Impact of a Mobile Information System on Changing Travel Behaviour and Improving Travel Experience. *Lecture Notes in Computer Science* 8093 LNCS: 2013, pp. 233–247.
2. Ibraeva, A., and J. F. De Sousa. Marketing of Public Transport and Public Transport Information Provision. *Procedia - Social and Behavioral Sciences* Vol. 162, 2014, pp. 121–128.
3. Kenyon, S., and G. Lyons. The Value of Integrated Multimodal Traveller Information and its Potential Contribution to Modal Change. *Transportation Research Part F: Traffic Psychology and Behaviour* Vol. 6, 2003, pp. 1–21.
4. Li, Z., D. A. Hensher, and J. M. Rose. Willingness to Pay for Travel Time Reliability in Passenger Transport: A Review and Some New Empirical Evidence. *Transportation Research Part E: Logistics and Transportation Review*, Vol. 46, 2010, pp. 384–403.
5. Toledo, T., and R. Beinhaker. Evaluation of the Potential Benefits of Advanced Traveler Information Systems. *Journal of Intelligent Transportation Systems*, Vol. 10, 2006, pp. 173–183.
6. Fosgerau, M., K. Hjorth, C. Brems, and D. Fukuda. *Travel Time Variability Definition and Valuation*. Technical Report. 2008.
7. Dziekan, K., and K. Kottenhoff. Dynamic At-Stop Real-Time Information Displays for Public Transport: Effects on Customers. *World Transit Research*, 2007.
8. Hou, X., and X. Chen. Analysis of Urban Public Transit Information Requirements in China by Web-based Survey. *Procedia - Social and Behavioral Sciences*, Vol. 96, 2013, pp. 1522–1527.
9. Brakewood, C., S. Barbeau, and K. Watkins. An Experiment Evaluating the Impacts of Real-Time Transit Information on Bus Riders in Tampa, Florida. *Transportation Research Part A: Policy and Practice*, Vol. 69, 2014, pp. 409–422.
10. Dyrberg, M. B., C. B. Christensen, M. K. Anderson, O. A. Nielsen, and C. G. Prato. Transfer Attributes in Route Choice Models for Public Transport Passengers. *Proc., 4th Symposium Arranged by European Association for Research in Transportation (hEART)*, Copenhagen, 2015.
11. Navarro, K. F., V. Gay, L. Golliard, B. Johnston, P. Leijdekkers, E. Vaughan, X. Wang, and M.-A. Williams. SocialCycle: What Can a Mobile App Do to Encourage Cycling? *Proc., 2013 IEEE 38th Conference on Local Computer Networks Workshops (LCN Workshops)*, IEEE, 2013, pp. 24–30.
12. Zografos, K. G., K. N. Androutsopoulos, and E. Apospori. User Acceptance and Willingness to Pay for the Use of Multimodal Trip Planning Systems. *Procedia-Social and Behavioral Sciences*, Vol. 48, 2012, pp. 2405–2414, 2012.
13. Molin, E. J. E., and H. J. P. Timmermans. Traveler Expectations and Willingness-to-Pay for Web-Enabled Public

Transport Information Services. *Transportation Research Part C: Emerging Technologies*, Vol. 14, 2006, pp. 57–67.
14. Dantan, S., J. Bulteau, and I. Nicolaï. Hurried or Risk-adverse Travelers: Who will Pay More for Multimodal Information? *Procedia Engineering*, Vol. 115, 2015, pp. 61–68.
15. Khattak, A. J., Y. Yim, and L. S. Prokopy. Willingness to Pay for Travel Information. *Transportation Research Part C: Emerging Technologies*, Vol. 11, 2003, pp. 137–159.
16. Yang, H. C. Bon Appétit for Apps: Young American Consumers' Acceptance of Mobile Applications. *Journal of Computer Information Systems*, Vol. 53, 2013, pp. 85–96.
17. Beul-Leusmann, S., C. Samsel, M. Wiederhold, K. H. Krempels, E. M. Jakobs, and M. Ziefle. Usability Evaluation of Mobile Passenger Information Systems. *Proc., International Conference of Design, User Experience, and Usability*, Springer, Cham, 2014, pp. 217–228
18. Pugesek, B. H., A. Tomer, and A. V. Eye, eds. *Structural Equation Modeling: Applications in Ecological and Evolutionary Biology*. Cambridge University Press, 2003.
19. Chen, C.-F., and W.-H. Chao. Habitual or reasoned? Using the Theory of Planned Behavior, Technology Acceptance Model, and Habit to Examine Switching Intentions Toward Public Transit. *Transportation Research Part F: Traffic Psychology and Behaviour*, Vol. 14, 2011, pp. 128–137.
20. Kaplan, S., J. de Abreu e Silva, and F. Di Ciommo. The Relationship Between Young People's Transit Use and Their Perceptions of Equity Concepts in Transit Service Provision. *Transport Policy*, Vol. 36, 2014, pp. 79–87.
21. Kaplan, S., M. Monteiro, M. K. Anderson, O. A. Nielsen, and E. Medeiros Dos Santos. The Role of Information Systems in Non-Routine Transit Use of University Students: Evidence from Brazil and Denmark. *Transportation Research Part A: Policy and Practice*, Vol. 95, 2017, pp. 34–48.
22. Golob, T. F. Structural Equation Modelling for Travel Behaviour Research. *Transportation Research Part B: Methodological*, Vol. 37, 2003, pp. 1–25.
23. Yuan, K. –H., and P. M. Bentler. Inferences on Correlation Coefficients in Some Classes of Nonnormal Distributions. *Journal of Multivariate Analysis*, Vol. 72, 2000, pp. 230–248.
24. White, H. A Heteroskedasticity-Consistent Covariance Matrix Estimator and a Direct Test for Heteroskedasticity. *Econometrica: Journal of the Econometric Society*, Vol. 48, 1980, pp. 817.
25. Bentler, P. M. *EQS 6 Structural Equations Program Manual.*
26. Browne, M. W., and R. Cudeck. Alternative Ways of Assessing Model Fit. *Sociological Methods & Research*, Vol. 21, 1992, pp. 230–258.
27. Hernandez, S., A. Monzon, and R. de Oña. Urban Transport Interchanges: A Methodology for Evaluating Perceived Quality. *Transportation Research Part A: Policy and Practice*, Vol. 84, 2016, pp. 31–43.
28. Ajzen, I. The Theory of Planned Behavior. *Organizational Behavior and Human Decision Processes*, Vol. 50, 1991, pp. 179–211.
29. Babbie, E. R. *The Practice of Social Research*. 12th ed. Wadsworth Cengage, Belmont, CA, London, 2010
30. Ullman, J. B., P. M. Bentler, J. B. Ullman, and P. M. Bentler. Structural Equation Modeling. In: *Handbook of Psychology*. John Wiley & Sons, Inc., Hoboken, NJ, USA, 2003
31. Loehlin, J. C., and A. A. Beaujean. *Latent Variable Models: An Introduction to Factor, Path, and Structural Equation Analysis*. 4th ed. Erlbaum, Mahwah, NJ, 2004.

The Standing Committee on Effects of Information and Communication Technologies (ICT) on Travel Choices (ADB20) peer-reviewed this paper (18-04688).

Article

TRR
JOURNAL OF THE TRANSPORTATION RESEARCH BOARD

Transportation Research Record
2018, Vol. 2672(50) 66–78

DOI: 10.1177/0361198118798460
journals.sagepub.com/home/trr
SAGE

Wired at Birth: Childhood, Technology Engagement, and Travel Behavior

Lama BouMjahed[1] and Hani S. Mahmassani[2]

Abstract

The ongoing demographic transition from baby boomers to millennials and technology-fueled evolution of transportation bring to the fore key trends that will determine how the future of transportation systems will unfold. This paper examines the implications of such trends on travel behavior by quantifying the impacts of a technologically engaged childhood—generally attributed to millennials—and an increased use of technology during travel on travel behavior now and in the future. Results indicate that individuals who grew up with a stronger childhood technology experience are more likely to be technologically engaged as adults, have stronger pro-environmental attitudes, lower car dependence, and stronger interest in autonomous vehicles. The results also support a simultaneous relationship between technology use during travel and car dependence, which posits that individuals with lower car dependence (or availability) are more likely to use technology during travel. Emanating results put forward potential benefits of policy starting at the childhood level. The implications of this work will only gain importance as the interplay between technology and travel deepens and as a larger share of the adult population become "digital natives."

The ongoing demographic transition and technology-fueled evolution of transportation bring to the fore key trends that will determine how the future of transportation systems will unfold. Indeed, young adults' increased differentiation from older generations (*1*) and their corresponding significant technology engagement (*2*) raise important questions for research. The emphasis on young adults' travel comes from the realization that their decisions will determine the use of future transportation systems as their relative share of the adult population, and of overall transportation system users, increases (*3*).

Recently, much work has burgeoned to try to understand the extent to which young adults differ from previous generations (*1*) and whether they will continue to differ (*4*). A key theory posits that their different preferences and travel behavior could be attributed to their increased technology engagement (*5*). To test this hypothesis, this paper sets out first to examine the role of technology engagement on preferences and behavior in the population more broadly. Then, the paper sets out to capture how a technologically engaged childhood determines behavior during adulthood in order to dig deeper into the reasons behind millennials' (or a population subgroup brought up during major technology innovations) differentiation at a more fundamental level. The rationale underpinning such an undertaking is the attribution of much of millennials' behavior by research to cohort effects or the uniqueness of their generation in growing with major technology innovations (*6*). Finally, this paper sets out to understand how these trends could be extended to anticipate future behavior, such as the adoption of autonomous vehicles.

Focusing on car dependence, technology engagement within and outside of transportation, residential choice, and environmental attitudes, this paper leverages data from an online survey to answer questions such as: *How does a technologically-engaged childhood affect adulthood preferences and behavior? What role does adulthood technology engagement play in preferences and activity-travel behavior? What does it mean for the future?* Answers to such questions can provide insights into the relevance of cohort effects and the importance of including such effects in research and planning that will only gain importance as the interplay between technology and travel grows and as a larger share of the population become "digital natives."

After a description of the background to place the study in its relevant context, a conceptual framework is

[1]Department of Civil Engineering, Northwestern University, Evanston, IL
[2]Transportation Center, Northwestern University, Evanston, IL

Corresponding Author:
Address correspondence to Hani S. Mahmassani:
masmah@northwestern.edu

developed. Thereafter, the methodology used is described, including the behavioral survey designed to assess the link between childhood and adulthood travel behavior. Next, the main results are presented, followed by a conclusion describing research implications.

Background

Young Adults and Cohort Effects

Young adults appear to exhibit signs of differentiation relative to older generations, such as a more significant technology engagement (*2*), changing preferences toward the environment (*7*) and urban settings (*8*), and a movement away from cars (*9*).

Several researchers used data from the National Household Travel Survey to suggest that, as a generation, young adults drive less than previous generations did at the same age (*1, 9*). This trend was also mirrored in the acquisition rate of driver's licenses. According to Sivak and Schoettle (*10*), less than half of 16-year-olds had a driver's license in 2016 compared to 80% in the 1980s. As a generation, they have been shown to use, to a greater extent, alternative modes such as public transit (*11*), active travel (such as walking or biking) (*12*), and on-demand mobility (*7*) and to be more multimodal than older generations (*13*).

A sizeable body of research has sought to understand whether these observed differences are structural or transient. Researchers have put forward three hypotheses: *cohort effects*, or the generation as a whole being intrinsically different from older generations (*9*); *period effects*, or having come of age at the onset of a financial crisis which drives their preferences and behavior (*1*); and *life-cycle effects*, or having yet to reach major milestones that require them to feel or act in a certain way (*1, 14*).

The cohort effect argument relies conceptually on the generation's "digital nativeness" (*15*), or the uniqueness of this generation in never having known a world without technology and its impact on adulthood travel. However, only one work in the literature, to the authors' knowledge, has directly linked the use of technology during childhood on future uses and behavior (*16*). Further, that work focuses more on the life-cycle uses rather than on capturing the childhood experience with technology as transmitted by socialization agents retrospectively. This is despite successful work in travel behavior measuring childhood experience retrospectively and its impact on adulthood travel determinants (*17–20*). This paper tackles this limitation.

Increased Role of ICT in Preferences and Activity-Travel Behavior

Within cohort effects, much of young adults' movement away from cars has been attributed to their increased use of technology. This link, between technology engagement and moving away from cars, relies on a more established field of research linking the use of information and communication technologies (ICT) to activities and travel behavior (*21, 22*) that evolved from examining the role of ICT on travel more broadly to an investigation of the role of technology in preferences and a disaggregated travel process.

In general, the literature suggests that ICT can have either a *substitution effect* on travel, where physical activities are replaced by virtual activities, or a *complementary effect*, where incremental travel is generated by the use of ICT (*21*). As an example, Astroza et al. (*23*) showed empirically that smartphone ownership and use is associated with the use of multiple modes of transportation, the pursuit of more complex tours with a larger number of stops per tour, engagement in recreational tours, and participation in joint tours with an accompanying person.

More recently, the link between ICT and preferences has gained prominence in the literature. Frei et al. (*24*) showed that ICT can alter attitudes toward travel by making time on-board more productive. On the other hand, Astroza et al. (*23*) used the 2014–2015 Puget Sound Regional Travel Study to define a green lifestyle construct to find that technology-savvy individuals were more likely to favor a green lifestyle and vice versa. Similar results were found by Wu et al.'s (*16*) analysis of the British Household Panel Survey.

Within this discussion, the literature remains limited in terms of the way technology interaction is integrated with travel decision-making processes. Conceptually, Mokhtarian and Tal (*25*) discussed the impact of ICT on the travel process and its role in trip generation, destination choice, mode choice, the evaluation of alternative modes, accessing modes (i.e., practical form for ride share matching) and facilitating travel (e.g., productivity during travel). Empirically, researchers have examined how travelers use ICT to obtain information on travel alternatives and destinations (*23*), and how they use it to facilitate travel (*7*) (or both). Yet, largely, no work has looked at ICT use in preliminary steps (such as the inception of the need to travel) or posterior steps (such as reflection over travel and impact on future behavior). This paper tackles this limitation.

Impacts on Adoption of Autonomous Vehicles

While the two trends outlined previously have stark implications for the future of urban travel, so does the market penetration and adoption of autonomous vehicles (AV). The literature has been relatively limited in exploring how the trends defining cohort effects and ICT might determine AV adoption. Most research is focused instead on understanding the benefits of its implementation. For example, Fagnant and Kockelman (*26*) argue that this new technology has the potential to "reduce crashes, ease congestion, improve fuel economy, reduce parking needs, bring mobility to those unable to drive, and over time dramatically change the nature of U.S. travel." Lavasani et al. (*27*) examined the willingness to pay for AV adoption and the likelihood of residential relocation based on a preference survey conducted in Miami, Florida. Their results showed that individual and household attributes, daily commute characteristics, demand for parking, and relative perceptions of AV were important factors in the decision. Only recently have Lavieri et al. (*28*) examined elements of tech savviness and their link to the use of autonomous vehicles to show that younger urban residents who are more educated and tech-savvy are more likely to be early adopters of AVs. This paper addresses the limitation in the literature by leveraging findings in terms of childhood and adulthood ICT impacts on preferences and travel while examining any potential effects these may have on AV adoption going forward.

Conceptual Framework

The framework in Figure 1 first integrates the cohort–age–period effects discussion by recognizing, in addition to the childhood experience (i.e., the root of cohort effects), life-cycle events (i.e., age/life-cycle effects) and period conditions (reflected in both economic conditions and emerging transportation solutions). It then extends into a human life-stage approach by recognizing the need to anticipate future behavior, such as the adoption of autonomous vehicles. The double-sided arrows are meant to highlight the importance of considering a typology of relationships between technology engagement, preferences, and activity-travel behavior. Travel behavior was disaggregated to allow for measurement of ICT interactions along the entire travel process from identifying the need for travel (e.g., discovering an event on Facebook), to learning about modes (looking for information and comparing modes), to accessing/acquiring modes (e.g., using a smartphone to hail a cab), to using modes (e.g., navigating using a GPS unit), and finally, reflecting on the modes after travel (e.g., rating the service using the mobile app of the service provider). This disaggregated framework is adapted from the EKB model (*29*) in consumer behavior.

Below, we specify a structural equation model to apply this framework to car dependence and interest in future autonomous vehicles.

Methodology

Procedure and Participants

The work in this paper is based on an online survey that was distributed via the commercial platform Amazon Mechanical Turk (MTurk) in February–March 2017 to U.S. residents. Mturk is a platform that connects requesters (i.e., researchers) with over 100,000 workers (i.e., respondents) from over 100 countries by offering an integrated recruiting and payment platform (*30*). MTurk participants are slightly more demographically diverse than standard internet samples and are significantly more

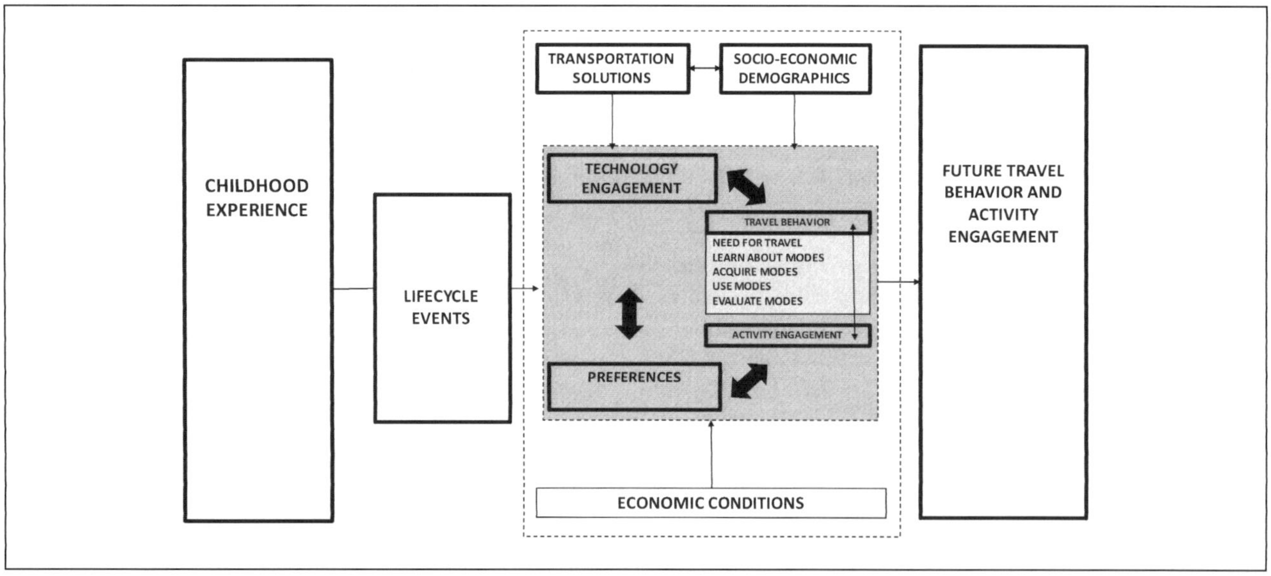

Figure 1. A life-stage approach conceptualization of young adults' activity-travel behavior.

diverse than typical U.S. college samples (*30*). Before releasing the survey on MTurk, it was tested on a convenience sample to capture possible survey fatigue and language ambiguity and was modified accordingly. MTurk links respondents to SurveyMonkey.com, which hosts the survey.

The survey sections stem directly from the described conceptual framework. Aside from a segment on sociodemographics, it included sections on childhood factors, life-cycle events, measures of technology engagement, preferences, and activity-travel behavior. It also incorporated questions on the interest in autonomous vehicles (personal and shared) when they become available. For childhood factors, respondents were asked to express their level of agreement or disagreement on a five-point Likert scale with statements capturing their extent of use of ICT during childhood. The contents of the questions were based on the theory of childhood socialization (*31*), highlighting the role of family, school, and peer groups in "transmitting the culture of a society" that persists with individuals until adulthood. Parents' attitude and behavior have also been found to impact chidlren's travel in fields outside of travel socialization (*32*).

For technology engagement, respondents were asked to express on a five-point Likert scale their frequency of use of and reliance on a smartphone for different activities (such as leisure, work, maintenance). Their daily media use was measured using scales borrowed from the literature (*33*). They were also asked to express their frequency of smartphone use throughout the travel process on a five-point Likert scale. Attitudes toward residential location and the environment were captured using statements borrowed from Tian et al. (*34*), Bagley and Mokhtarian (*35*), and Jariyasunant et al. (*36*). Finally, travel behavior was measured by asking respondents to self-report the number of trips, by modes and activities, made in the previous week. In addition, they were asked to report the number of days on which they engaged in virtual activities in the past week instead of outside the home.

The resulting data were pre-processed to (sequentially) remove incomplete responses (11%), flatliners (i.e., respondents who complete entire sections ostensibly without looking at the questions; 11%), speeders (i.e., respondents who finish the survey in an unusually short amount of time; 1.5%), and those who failed internal consistency checks (1%). Individuals with incomplete responses and flatliners were not paid. In all, 932 complete responses were used in the analysis presented throughout this paper. Their characteristics are summarized in Table 1 and compared with U.S. statistics. The table comparison shows that the sample of this study is skewed toward females, millennials, car-oriented individuals, and less urban residents. No survey weights were used to reflect skewness, but it was recognized in the limitation of the work.

Measures

To study the relationships defined in our framework, we measured exogenous (Table 2) and endogenous (latent) variables (Table 3). In a cross-sectional setting, full-time employment was used as a proxy for economic conditions reflecting a potential lower impact of a financial crisis on individuals who have a (stable) full-time employment. Transportation solutions are captured by the ownership of a transit pass and the number of household cars. The walk score of the residential location was obtained through walkscore.com (*43*), a website that assigns a walkability number between 0 and 100 for any address in the United States. An API, associated with the WalkScore website within the R environment, was used to obtain the walk scores (*43*). The "car-dependent" walk score (V3) used in this paper represents a walk score between 0 and 49 reflecting that there is "minimal to some transit" and that "most to all errands require a car." Self-reported zip codes were also used for the place of residence during childhood and mapped to an urban measure using the rural–urban continuum classification of the U.S. Department of Agriculture (*44*).

A factor analysis was performed for latent variables with "Promax" rotation for technology engagement, preferences/attitudes and the childhood experience. One latent variable was kept for the childhood technology experience, two for preferences/attitudes, and three for technology engagement. In each of the latent variables the indicators with the highest loading factors were included. In addition, two latent variables were created: one for car dependence and one for the interest in autonomous vehicles. The calculated Cronbach's alpha (*45*) for latent variables' internal consistency were estimated and included in Table 3 above. Two constructs have lower alphas than the 0.7 recommended as a rule of thumb by Nunnally (*46*); however, based on the nature of cutoffs for fit indices and previous research, it was decided to leave them but keep it in mind during analysis.

Analysis Approach

A structural equation model (SEM) was estimated using R software (*47*). It is a multivariate regression model in which the same variable can appear in more than one equation, as a predictor variable in some and a response variable in others (*48*). This is the case for many variables in our model (e.g., ENV).

The specification of the model was derived directly from the conceptual framework. First, car dependence and AV regression equations included elements of

Table 1. Survey Population Characteristics and Representativeness

N = 932	Share of total survey sample	Total U.S. adult population (18+)
Socio-demographics		
Female	59%	51%
Age/generation[a]		(36)
Silent generation (DOB 1928–1945)	1%	11%
Baby boomers (DOB 1946–1964)	14%	30%
Generation X (DOB 1965–1980)	33%	26%
Generation Y – millennials (DOB 1981–1997)	52%	30%
of which: old millennials (DOB 1981–1989)	57%	52%
of which: young millennials (DOB 1990–1997)	43%	48%
Generation Z (DOB 1998–)	<1%	2%
Adult household size		
1	20%	28%
2	58%	34%
3–5 +	22%	38%
Average household size	2.1	2.5
Economic characteristics		(37)
Occupation		
Student	4%	8%
Employed outside the home (35 + hours per week)	57%	45%[b]
Employed outside the home (<35 hours per week)	11%	15%[b]
Work from home/homemaker	18%	10%[c]
Unemployed, looking for work	3%	3%
Unemployed, not looking for work	2%	0.2%
Retired (NCES 2016)	3%	16%
Disabled/unable to work	2%	0.1%[d]
Annual income in US$		(36)
Less than $25,000	19%	21%
$25,000 + – $40,000	17%	14%
$40,000 + – $55,000	19%	12%
$55,000 + – $70,000	12%	10%
$70,000 + – $85,000	9%	9%
More than $85,000	23%	34%
Place of residence characteristics		
Urban	77%	81% (38)
Transportation characteristics		
Driver's license	95%	85% (39)
Owns a car	91%	91%[e]
Owns a bike	46%	53% (40)
Has a transit pass	11%	N/A
Participates in a car-share program	4%	0.5% (41)

[a]Percentages for a given category may not add up to 100% due to rounding.
[b]BLS data does not include location of employment.
[c]Working at home includes any time persons did work at home and is not restricted to persons whose usual workplace is their home. No data on homemakers.
[d]Persons having claimed ill health or disability as a reason for being outside of the labor force.
[e]Household car ownership (at least one car).

childhood and adulthood technology engagement, preferences/attitudes, life-cycle events, economic conditions, and socio-economic demographics. Second, the childhood technology engagement was included as an explanatory variable in all latent variables. Third, in addition to childhood, gender, age, land use, parenting status, and economic conditions were all included as explanatory variables for latent variables. Other potential variables could include mobility and activity engagement, but those were restricted to zero due to the limit on the number of parameters to be estimated with our sample size. Fourth, travel inertia was regressed on latent variables including a travel aspect. Finally, spillover effects from virtual socialization (SOC) to virtual information and communication (VIC) and the use of technology during travel (TECHTRAV) were specified reflecting a potential media dependency element captured in the "reliance on smartphone" element captured in SOC.

Table 2. Exogenous Variables Used in the Study

Variable	Description	Mean	SD
Age	V1: continuous (min = 18, max =85)	38	12
Number of cars in the household	V2: continuous (=0 to 5; with 5 for >=5)	1.8	1.0
Walk score of residential location	V3: binary (=1 if low walk score; i.e., car-dependent area as defined by the source)	0.78	–
Land use of the residential place during high school	V4: binary (=1 if Urban)	0.85	–
Land use of the current residential place	V5: binary (=1 if Urban)	0.77	–
Employed outside the house for 35 hours or more	V6: binary	0.57	–
Income	V7: continuous (min = 5k-max = 100k)	56,204	28,184
Number of days on which engaged in virtual "needs" activities	V8: continuous (min = 0, max = 7)	2.44	1.98
Travel inertia	V9: binary (=1 if the respondent would not change travel behavior when receiving an extra 100$/month to spend on transportation)	0.68	–
Transit pass ownership	V10: binary	0.11	–
Childbirth	V11: binary (=1 if experienced childbirth in the past 3 years)	0.15	–
Gender	V12: binary (=1 if Female)	0.59	–

Note: – = the standard deviation (SD) is not reported because it corresponds to a binary variable.

Another spillover was specified between the general use of technology (VIC) and technology use during travel.

To address non-normality of the data, the equations were estimated using maximum likelihood with robust standard errors and a Satorra-Bentler (*49*) scaled test statistic. Satorra and Bentler (*50*) developed a statistic that incorporates a scaling correction for the χ^2 statistic when distributional assumptions are violated. Its computation takes into account the model, the estimation method, and the sample kurtosis values. The corrected χ^2 statistic has been shown to be the most reliable test statistic for evaluating mean and covariance structure models under various distributions and sample sizes (*51*). In addition to chi-square, the literature recommends using both absolute and incremental fit indices such as the comparative fit index (CFI) and the root mean squared error of approximation (RMSEA). The calculated Robust RMSEA and CFI (*52*) included in the next section identify a good overall fit for the model.

Results

The estimated structural model (Figure 2) and the factor loadings of the measurement model (Table 4) are discussed below.

The main findings resulting from the model are summarized in Table 5 and discussed thereafter.

For the sake of brevity, we will discuss below the main relationships of interest, including childhood technology engagement, car dependence, and interest in autonomous vehicles. For other relationships, the reader can refer to Figure 2 and Table 5.

Childhood Technology Engagement

Two important results for childhood technology engagement are worth discussing. First, Figure 2 depicts a negative relationship between age (V1) and CHILD, suggesting that younger adults had a stronger technology engagement during childhood. Urban residence during childhood was not found to be statistically significant in its impact on CHILD. Second, the results in Table 5 (first row) suggest that the childhood technology construct (CHILD) affects adulthood technology engagement, preferences, travel behavior, and even interest in autonomous vehicles. Indeed, users who had a stronger technology engagement experience during childhood are more likely to: (1) use their smartphone for socialization (SOC); (2) spend increasing time on VIC; (3) be more pro-environmental (ENV); (4) be less car dependent as adults (CARDEP); and (5) show more interest in autonomous vehicles when they become available (AV). Younger individuals were also more likely to be technologically engaged (SOC and VIC). These results support the attribution of an increased use of technology during adulthood to a strong childhood technology experience. They could also be indicating a significant impact of technology during childhood on future preferences such as being pro-environmental. Such a result, if reproduced and confirmed, would encourage policy makers to start intervening at the childhood level to promote stronger pro-environmental attitudes during adulthood. It is worth noting that the relationship between childhood technology engagement and the use of technology during travel is not statistically significant—unlike age—suggesting that this behavior may be driven by elements of

Table 3. Latent Variables and Their Indicators

Variable description		Mean	SD	Alpha
(CARDEP) Car dependence				0.7
CAR1	The percentage of trips done by private personal car in the past week	0.68	0.34	
CAR2	Car availability[a]	4.4	1.13	
(TECHCHILD) Pro-technology childhood environment				0.8
CH1	ICT were an integral part of the quality time I spent with friends and family[b]	2.36	1.27	
CH2	I rarely used ICT[b]	3.33	1.43	
CH3	ICT were an integral part of my education at school[b]	2.42	1.25	
(SOC) Reliant on smartphone for socialization				0.9
SO1	Frequency of use of social networking-related apps (e.g., Facebook)[c]	4.03	1.42	
SO2	Reliance on smartphone for social networking (e.g., Facebook)[d]	3.57	1.25	
(VIC) Virtual information and communication				0.7
ME1	Time spent daily instant messaging/chatting[e]	1.83	1.50	
ME2	Time spent daily voice- or video-calling[e]	1.05	1.21	
ME3	Time spent daily online shopping (browsing, purchasing, rating, etc.)[e]	1.56	1.19	
ME4	Time spent daily trip-planning (e.g., Google Maps)[e]	1.13	1.07	
(ACCESS) Pro-access				0.6
AC1	Importance of desire for nearby shops and services in choosing the place of residence[f]	3.06	1.15	
AC2	Importance of access to freeways in choosing the place of residence[f]	2.56	1.18	
AC3	Importance of parking availability and cost in choosing the place of residence[f]	2.84	1.49	
(ENV) Pro-environmental				0.7
EN1	I value the benefits to society when I take sustainable modes of transportation[b]	3.53	0.99	
EN2	I feel guilty if I don't take sustainable modes of transportation[b]	2.40	1.11	
EN3	I wouldn't change my transportation behavior if it were only for the benefit of the environment[b]	2.95	1.11	
(TECHTRAV) Technology-enabled travel behavior				0.9
TT1	Frequency of using smartphone applications to identify possible destinations (e.g., café, restaurant)[g]	3.32	1.04	
TT1	Frequency of using smartphone applications to look for information to decide how to travel to your destinations[g]	3.29	1.11	
TT2	Frequency of using smartphone applications to compare travel alternatives[g]	2.82	1.16	
TT3	Frequency of using smartphone applications to access/acquire travel[g] alternatives (e.g., track your alternative, request your alternative, buy tickets for your alternative)[g]	2.51	1.16	
TT4	Frequency of using smartphone applications to facilitate travel (e.g., navigation, real-time coordination, transfer to other alternatives)[g]	3.20	1.19	
TT5	Frequency of using smartphone applications to evaluate/rate travel alternatives[g]	2.53	1.20	
(AVS) Interest in using autonomous vehicle services				0.8
AV1	Level of interest in autonomous vehicles when they become available in the future[h]	3.00	1.44	
AV2	Level of interest in shared autonomous vehicles (such as autonomous on-demand services like Uber) when they become available in the future[h]	2.71	1.40	

Note: L.S. = Likert scale; SD = standard deviation.
[a]5-point L.S. from nonexistent to unrestricted.
[b]5-point L.S. from strongly disagree to strongly agree.
[c]5-point L.S. from I don't have it on my phone to I have it on my phone and I frequently use it.
[d]5-point L.S. from never use to cannot live without.
[e]8-point L.S. from no time to more than 8 hrs.
[f]5-point L.S. from not at all important to extremely important.
[g]5-point L.S. from never to always.
[h]5-point L.S. from not at all interested to very interested.

age effects rather than cohort effects attributed to the childhood experience.

Car Dependence

On the other hand, the results (first column) support the framework conceptualization suggesting that cohort effects, period effects, life-cycle effects, travel inertia, technology engagement, preferences and attitudes, and traditional socio-economic demographics (SED) and land use affect car dependence.

Several noteworthy relationships predict lower car dependence. Individuals that had significant technology engagement during childhood and adulthood were less

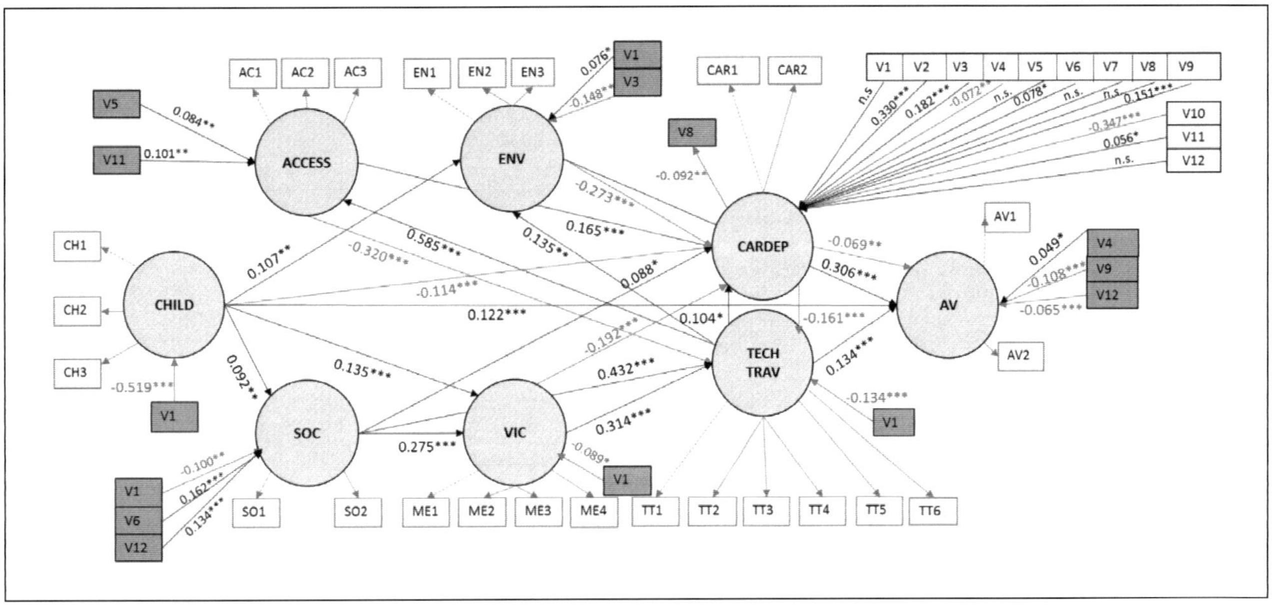

Figure 2. Schematization of the SEM.

Note: *N = 932 || CFI = 0.90 & RMSEA = 0.045 || chisq/df = 2.7 || N/obs = 16 || $p < 0.1$, $p < 0.5$**, $p < 0.01$***;
Rectangles represent observed variables and circles represent latent variables; V1 = age; V2 = number of household cars; V3 = car-dependent adulthood place of residence; V4 = urban place of residence during childhood; V5 = current urban place of residence; V6 = full-time employment; V7 = income; V8 = needs virtual activities; V9 = travel inertia; V10 = transit pass; V11 = childbirth; V12 = gender (female). The directed arrows describe the regression from the origin to the destination. The coefficient of the arrows represents the strength of the relationship. The dashed arrows represent the indicators used to set the scale of the latent variable.

likely to be car dependent. The latter echoes findings in Lavieri et al. (*53*). Similarly, individuals who grew up in an urban environment during high school were predicted to be less car dependent. Finally, pro-environmental (ENV) individuals and those who own a transit pass were less likely to be car dependent.

Conversely, several relationships illustrate key drivers of higher car dependence. Individuals who work full-time (V5) or have had a child in the past three years (V9) were linked to higher car dependence, potentially reflecting the need for private cars. Individuals who value access to parking, freeways, and shops (ACCESS), have stronger travel inertia (V8), live in car-dependent environments (V3), or in a household with a larger number of cars are more likely to be car dependent. The urban environment of the current place of residence, age, gender, and income were insignificant in the model, despite research showing the statistical significance of all three on car dependence.

Car dependence was also tested as a predictor variable for the use of technology for travel and engagement in virtual activities (V8). The simultaneous relationships showed that individuals who use more technology during travel are more car dependent while those who have a lower car dependence (i.e., lower car availability) are more likely to use technology throughout travel (e.g., Uber). The former relationship is less intuitive, perhaps suggesting that the use of technology during travel is enabling private car travel (e.g., navigating to unfamiliar locations). On the other hand, there was no statistical significance of the impact of virtual activity engagement on car dependence but rather a relationship indicating that people who are less car dependent (i.e., lower car availability) are more likely to engage in virtual activities (e.g., online banking).

However, this work remains exploratory and provides useful hypotheses for future work. Specifically, it would be important to consider investigating the impact of car dependence on preferences and technology use, the choice of residential location (i.e., self-selection), and running more equivalent models to establish causality (*54*).

Other Relationships

The model explored several interesting relationships within technology engagement and preferences/attitudes (middle-top part of Table 5). First, the model supported the hypothesized media dependency relationship (*55*) suggesting that individuals who depend on their smartphones for socialization are more likely to use it for travel and the spillover from VIC to TECHTRAV. Second, the model indicated that technology engagement

Table 4. Measurement Model Factor Loadings

Variable description		Factor loading
(CARDEP) Car dependence		
CAR1	The percentage of trips done by private personal car in the past week	1.000
CAR2	Car availability	3.063
(CHILD) Pro-technology childhood experience		
CH1	ICT were an integral part of the quality time I spent with friends and family	1.000
CH2	I rarely used ICT[a]	0.953
CH3	ICT were an integral part of my education at school	0.769
(SOC) Smartphone use for socialization		
SO1	Frequency of use of social networking-related apps (e.g., Facebook)	1.000
SO2	Reliance on smartphone for social networking (e.g., Facebook)	0.991
(VIC) Virtual information and communication		
ME1	Time spent daily instant messaging/chatting	1.000
ME2	Time spent daily voice- or video-calling	0.861
ME3	Time spent daily online shopping (browsing, purchasing, rating, etc.)	0.792
ME4	Time spent daily trip-planning (e.g., Google Maps)	0.848
(ACCESS) Pro-access		
AC1	Importance of desire for nearby shops and services in choosing the place of residence	1.000
AC2	Importance of access to freeways in choosing the place of residence	1.543
AC3	Importance of parking availability and cost in choosing the place of residence	1.194
(ENV) Pro-environmental		
EN1	I value the benefits to society when I take sustainable modes of transportation (e.g., public transportation, walking/biking, etc.)	1.000
EN2	I feel guilty if I don't take sustainable modes of transportation (e.g., public transportation, walking/biking, etc.)	0.934
EN3	I wouldn't change my transportation behavior if it were only for the benefit of the environment[a]	0.769
(TECHTRAV) Technology-enabled travel behavior		
TT1	Frequency of using smartphone applications to identify possible destinations (e.g., café, restaurant)	1.000
TT2	Frequency of using smartphone applications to look for information to decide how to travel to your destinations	1.109
TT3	Frequency of using smartphone applications to compare travel alternatives	1.278
TT4	Frequency of using smartphone applications to access/acquire travel alternatives (e.g., track your alternative, request your alternative, buy tickets for your alternative)	1.206
TT5	Frequency of using smartphone applications to facilitate travel (e.g., navigation, real-time coordination, transfer to other alternatives)	1.140
TT6	Frequency of using smartphone applications to evaluate/rate travel alternatives	1.193
(AV) Interest in using autonomous vehicle services		
AV1	Level of interest in autonomous vehicles when they become available in the future	1.000
AV2	Level of interest in shared autonomous vehicles (such as autonomous on-demand services like Uber) when they become available in the future	0.768

[a]Coded in reverse order from the rest of the indicators within the latent variables.

affects preferences/attitudes where TECHTRAV predicted a stronger preference for access in the place of residence and pro-environmental attitudes. The relationship between technology engagement and pro-environmental attitudes was also found in Lavieri et al. (*53*). As for access, it could mean that using technology for travel made people more sensitive to dimensions of time and distance and therefore aspects of access. The reverse relationship between TECHTRAV and ACCESS was also supported whereby individuals with a lower preference for access were more likely to use technology during travel, perhaps suggesting the need for navigation. Examination of alternative models for the relationships between technology engagement and preferences/attitudes may be helpful to further rule out the possibility of spurious correlations.

The Implications for Autonomous Vehicles

Finally, the model reveals interesting conclusions about who will be most interested in autonomous vehicles when they become available (last column). In addition to individuals who grew up with technology, male individuals who lived in an urban residence during childhood, are more pro-environmental, use more technology during travel, have stronger travel inertia, and are less car dependent show more interest in autonomous vehicles. The pro-environmental element is in line with Lavieri et al.'s (*28*) findings for shared autonomous vehicles. This provides insights into how services such as autonomous vehicles, if to be used as complements to sustainable modes (e.g., transit), should be marketed. Relatively speaking, growing up with technology had a stronger impact on

Table 5. Standardized Direct Effects of Model Variables

	CARDEP	SOC	VIC	ENV	ACCESS	TECHTRAV	AV
Variance explained	59%	7%	13%	11%	4%	13%	21%
CHILD	–0.114(0.008)***	0.092(0.045)**	0.135(0.037)***	0.107(0.034)*	n.s.	n.s.	0.122(0.040)***
SOC	0.088(0.009)*		0.275(0.029)***	n.s.	–	0.432(0.032)***	n.s.
VIC	–0.192(0.014)***	–		n.s.	–	0.314(0.047)***	n.s.
TECHTRAV	0.104(0.017)*	–	–	0.135(0.045)**	0.585(0.073)***		0.134(0.065)***
ENV	–0.273(0.016)***	–	–		–	–	0.306(0.083)***
ACCESS	0.165(0.021)***	–	–	–		–0.32(0.122)***	n.s.
CARDEP		–	–	–	–	–0.161(0.162)***	–0.069(0.227)**
Age (V1)	n.s.	–0.1(0.005)**	–0.089(0.004)*	0.076(0.003)*–	n.s.	–0.134(0.003)***	n.s.
HH Cars (V2)	0.330(0.011)***	–	–	–	–	–	n.s.
Car-dependent walk score (V3)	0.182(0.027)***	–	–	–0.148(0.063)***	–	–	n.s.
Urban high school residence (V4)	–0.072(0.024)*	n.s.	n.s.	n.s.	n.s.	n.s.	0.049(0.112)*
Urban current residence (V5)	n.s.	n.s.	n.s.	n.s.	0.084(0.057)**	n.s.	n.s.
Full-time employee (V6)	0.078(0.020)*	0.162(0.085)***	n.s.	–	n.s.	–	n.s.
Income (V7)	n.s.	–	–	–	–	–	n.s.
Virtual Needs (V8)	n.s.	–	–	–	–	–	n.s.
Travel inertia (V9)	0.151(0.022)***	–	–	–	–	n.s.	–0.108(0.098)***
Transit pass (V10)	–0.347(0.044)***	–	–	–	–	–	n.s.
Childbirth (V11)	0.056(0.023)*	n.s.	n.s.	–	0.101(0.065)**	n.s.	n.s
Gender (V12)	n.s.	0.134(0.078)***	n.s.	n.s.	n.s.	n.s.	–0.065(0.067)**

Note: (standard errors); n.s. = that the relationship was not significant at the 0.01 level and was dropped from the model; – = that the relationship was restricted to zero by design. *$p < 0.01$, **$p < 0.05$, **$p < 0.001$; the table does not include the CHILD and V8 equations.

interest in AVs than other SED variables. Still, pro-environmental attitudes and technology use during travel had the strongest impact on AVs.

Discussion and Limitations

Overall, these results put forward the role of childhood technology engagement on adulthood behavior. It also argues for the importance of addressing the link between increased technology use during travel and preferences and behavior. For car dependence, the results suggest that while individuals who grew up with stronger technology engagement might be less inclined to use or rely on private cars, it does not necessarily imply that as a generation they would move away from cars. This is because results point to the role of elements such as life cycle, employment, and the built environment, as well as preferences and attitudes in car dependence. In fact, while the impact of CHILD (−0.114) on CAR was stronger than childbirth (0.056) and employment (0.078), it is weaker than traditional socio-demographic attributes such as household cars (0.33) and transit pass (0.35) as indicated by the standardized coefficients in the CAR equation (column 1).

While the model presents interesting results, several limitations should be noted. The first pertains to the non-randomness and therefore non-representativeness of the data which hinders generalization efforts. Second, the nature of the survey imposes a limitation. Being "attitudinal" makes it susceptible to unintended wording biases. Moreover, an important section of the survey is "retrospective," where respondents are asked to recall or reflect on a situation that may be far away in time. This may engender memory biases in the responses. An example is the zip code during childhood. Also, we use proxies of economic conditions such as full-time employment, which could better be captured with panel data. Furthermore, the specification of the model is limited. For example, the model does not separate between autonomous vehicles and shared autonomous vehicles, an important distinction the literature suggests should be made (*28*). Also, the model does not capture geographic differences within the U.S. Finally, the explanatory power of the model (i.e., the variance explained) is relatively low. This is not unusual with disaggregate, individual-level data in areas where considerable heterogeneity is likely to be present. Nonetheless, the relationships described throughout are plausible and meet the study intent in light of the exploratory nature of the study.

Conclusion

The main motivation behind this work stems from urban sustainability and the critical role that transportation plays therein. The ongoing demographic transition and technology-fueled evolution of transportation bring to the fore key trends that will determine how the future of transportation systems will unfold. In that respect, this paper set out to understand how millennial-specific factors such as an increased childhood and adulthood technology engagement affect preferences and travel behavior with a focus on car dependence. It then set out to assess what these relationships mean for the future in terms of interest in autonomous vehicles.

This paper offers several contributions. To the best of the authors' knowledge, this paper is the first to capture (retrospectively) the technology experience during childhood, and its link to adulthood preferences and travel behavior. It is also the first to capture a technology engagement throughout a disaggregated travel process extending beyond information search and navigation. It also adds to the limited literature examining the interest in autonomous vehicles from a travel behavior perspective including attitudes and technology engagement. These are not major contributions but add useful insight to the relatively small body of work in this area, especially in the transportation community.

The findings presented throughout this paper are encouraging and lay the foundation for more extensive research by highlighting areas that could benefit from more in-depth examination. Several of the findings echo others found in the literature and support intuitive relationships, but other findings put forth noteworthy results. For example, the opportunity arises for policy makers to work toward sustainable travel solutions and potentially affect attitudes by intervening at the childhood level through giving equal ICT access to children. Such access could teach children to enable non-motorized modes by using ICT, thus creating long-term impacts. Understanding the advantages and challenges of the widespread adoption of autonomous vehicles is also critical for promoting sustainable travel, if used in combination with sustainable modes. The implications of the work presented in this paper will only gain importance as the interplay between technology and travel deepens and as a bigger share of the adult population become "digital natives."

Acknowledgment

Partial funding for this work is provided through a dissertation year fellowship to the first author by the Northwestern University Transportation Center.

Author Contributions

Both authors contributed to all aspects of the study from conception and design, to data collection, to analysis and interpretation of results, and manuscript preparation. Both authors reviewed the results and approved the final version of the manuscript.

References

1. Blumenberg, E., B. D. Taylor, M. Smart, K. Ralph, M. Wander, and S. Brumbagh. What's youth got to do with it? Exploring the travel behavior of teens and young adults. University of California Transportation Center, Berkeley, CA, 2012.
2. Peter, K. *Infographic: Millennials & Media Consumption*, October 2014.
3. Fry, R. *Millennials Overtake Baby Boomers as America's Largest Generation*. Pew Research Center, Washington, D.C., Vol. 25, 2016.
4. Garikapati, V. M., R. M. Pendyala, E. A. Morris, P. L. Mokhtarian, and N. C. McDonald. Activity Patterns, Time Use, and Travel of the Millennial Generation: What is All the Hype About? Presented at 95th Annual Meeting of the Transportation Research Board, Washington, D.C., 2016, 558–584.
5. van Wee, B. Peak Car: The First Signs of a Shift Towards ICT-based Activities Replacing Travel? A Discussion Paper. *Transport Policy*, Vol. 42, 2015, pp. 1–3.
6. Lyons, G. Transport's Digital Age Transition. *Journal of Transport and Land Use*, Vol. 8, No. 2, 2015, pp. 1–19.
7. Circella, G., L. Fulton, F. Alemi, R. M. Berliner, K. Tiedeman, P. L. Mokhtarian, and S. Handy. *What Affects Millennials' Mobility? PART I: Investigating the Environmental Concerns, Lifestyles, Mobility-Related Attitudes and Adoption of Technology of Young Adults in California*. Project Report. National Center for Sustainable Transportation, University of California, Davis, CA, 2016.
8. Center, T. *Who's On Board: 2014 Mobility Attitudes Survey*. Transit Center, New York City, NY, 2014.
9. McDonald, N. C. Are Millennials Really the "Go-nowhere" Generation? *Journal of the American Planning Association*, Vol. 81, No. 2, 2015, pp. 90–103.
10. Sivak, M., and B. Schoettle. Update: Percentage of Young Persons with a Driver's License Continues to Drop. *Traffic Injury Prevention*, Vol. 13, No. 4, 2012, pp. 341–341
11. Grimsrud, M., and A. El-Geneidy. Transit to Eternal Youth: Lifecycle and Generational Trends in Greater Montreal Public Transport Mode Share. *Transportation*, Vol. 41, No. 1, 2014, pp. 1–19.
12. Davis, B., T. Dutzik, and P. Baxandall. *Transportation and the New Generation: Why Young People are Driving Less and What it Means for Transportation Policy*. Frontier Group and U.S. PIRG Education Fund, Washington, D.C., 2012.
13. Ralph, K. M. Multimodal Millennials? The Four Traveler Types of Young People in the United States in 2009. *Journal of Planning Education and Research*, Vol. 37, No. 2, 2017, pp. 150–163.
14. Klein, N. J., and M. J. Smart. Millennials and Car Ownership: Less Money, Fewer Cars. *Transport Policy*, Vol. 53, 2017, pp. 20–29.
15. Prensky, M. Digital Natives, Digital Immigrants Part 1. *On the Horizon*, Vol. 9, No. 5, 2001, pp. 1–6.
16. Wu, G., J. Hong, and P. V. Thakuriah. The Role of the Internet in Shaping Young Adults' Attitude, Travel Choices and Sustainable Lifestyles: A Longitudinal Perspective. Presented at 95th Annual Meeting of the Transportation Research Board, Washington, D.C., 2017.
17. Haustein, S., C. A. Klöckner, and A. Blöbaum. Car Use of Young Adults: The Role of Travel Socialization. *Transportation Research Part F: Traffic Psychology and Behaviour*, Vol. 12, No. 2, 2009, pp. 168–178.
18. Mjahed, L. B., C. Frei, and H. S. Mahmassani. Walking Behavior: The Role of Childhood Travel Experience. *Transportation Research Record: Journal of the Transportation Research Board*, 2015. 2495: 94–100.
19. Underwood, S., and S. Handy. *Adolescent Attitudes Towards Active Transportation: Bicycling in Youth in Retrospect from Adulthood*. No UCD-ITS-RR-12-14. 2012.
20. Underwood, S. K., S. L. Handy, D. A. Paterniti, and A. E. Lee. Why do Teens Abandon Bicycling? A Retrospective Look at Attitudes and Behaviors. *Journal of Transport & Health*, Vol. 1, No. 1, 2014, pp. 17–24.
21. Choo, S., T. Lee, and P. Mokhtarian. Do Transportation and Communications Tend to be Substitutes, Complements, or Neither? US Consumer Expenditures Perspective, 1984-2002. *Transportation Research Record: Journal of the Transportation Research Board*, 2007. 2010: 121–132.
22. Cohen-Blankshtain, G., and O. Rotem-Mindali. Key Research Themes on ICT and Sustainable Urban Mobility. *International Journal of Sustainable Transportation*, Vol. 10, No. 1, 2016, pp. 9–17.
23. Astroza, S., V. M. Garikapati, C. R. Bhat, R. M. Pendyala, P. S. Lavieri, and F. F. Dias. Analysis of the Impact of Technology Use on Multi-Modality and Activity-Travel Characteristics. *Transportation Research Record, Journal of the Transportation Research Board*, 2017. 2666: 19–28.
24. Frei, C., H. S. Mahmassani, and A. Frei. Making Time Count: Traveler Activity Engagement on Urban Transit. *Transportation Research Part A: Policy and Practice*, Vol. 76, 2015, pp. 58–70.
25. Mokhtarian, P. L., and G. Tal. *Impacts of ICT on Travel Behavior: A Tapestry of Relationships*. Handbook of Transport Studies, SAGE Publications, Inc., Thousand Oaks, CA, 2013.
26. Fagnant, D. J., and K. Kockelman. Preparing a Nation for Autonomous Vehicles: Opportunities, Barriers and Policy Recommendations. *Transportation Research Part A: Policy and Practice*, Vol. 77, 2015, pp. 167–181.
27. Lavasani, M., H. Asgari, X. Jin, and A. Pinjari. Investigating the Willingness to Pay for Autonomous Vehicles and the Likelihood of Residential Relocation. Presented at 96th Annual Meeting of the Transportation Research Board, Washington, D.C., 2017.
28. Lavieri, P. S., V. M. Garikapati, C. R. Bhat, R. M. Pendyala, S. Astroza, and F. F. Dias. Modeling Individual Preferences for Ownership and Sharing of Autonomous Vehicle Technologies. *Transportation Research Record: Journal of the Transportation Research Board*, 2017. 2665: 1–10.
29. Kollat, D. T., J. F. Engel, and R. D. Blackwell. Current problems in consumer behavior research. *Journal of Marketing Research*, 1970, pp. 327–332.

30. Buhrmester, M., T. Kwang, and S. D. Gosling. Amazon's Mechanical Turk: A New Source of Inexpensive, Yet High-quality, Data? *Perspectives on Psychological Science*, Vol. 6, No. 1, 2011, pp. 3–5.
31. Baslington, H. Travel Socialization: A Social Theory of Travel Mode Behavior. *International Journal of Sustainable Transportation*, Vol. 2, No. 2, 2008, pp. 91–114.
32. Driller, B. K., and S. L. Handy. Exploring the Influence of Parents on Children's Bicycling in Davis, CA. Presented at 92nd Annual Meeting of the Transportation Research Board, Washington, D.C., 2013, 13–17.
33. Rosen, L. D., K. Whaling, S. Rab, L. M. Carrier, and N. A. Cheever. Facebook Creating "iDisorders"? The Link Between Clinical Symptoms of Psychiatric Disorders and Technology Use, Attitudes and Anxiety. *Computers in Human Behavior*, Vol. 29, No. 3, 2013, pp. 1243–1254.
34. Tian, G., R. Ewing, and W. Greene. Desire for Smart Growth: A Survey of Residential Preferences in the Salt Lake Region of Utah. *Housing Policy Debate*, Vol. 25, No. 3, 2015, pp. 446–462.
35. Bagley, M. N., and P. L. Mokhtarian. The Impact of Residential Neighborhood Type on Travel Behavior: A Structural Equations Modeling Approach. *The Annals of Regional Science*, Vol. 36, No. 2, 2002, pp. 279–297.
36. Jariyasunant, J., M. Abou-Zeid, A. Carrel, V. Ekambaram, D. Gaker, R. Sengupta, and J. L. Walker. Quantified Traveler: Travel Feedback Meets the Cloud to Change Behavior. *Journal of Intelligent Transportation Systems*, Vol. 19, No. 2, 2015, pp. 109–124.
37. Census. 2016a. https://www.census.gov/. Accessed October 20, 2016.
38. U.S. Bureau of Labor Statistics. 2016. https://www.bls.gov/news.release/atus.t06.htm. Accessed October 20, 2016.
39. Census. 2016b. https://www.census.gov/library/visualizations/2016/comm/acs-rural-urban.html. Accessed October 20, 2016.
40. Federal Highway Administration. 2015. https://www.fhwa.dot.gov/policyinformation/pubs/hf/pl11028/chapter4.cfm. Accessed October 20, 2016.
41. National Household Travel Survey. 2009. http://nhts.ornl.gov/. Accessed October 20, 2016.
42. Transportation Sustainability Research Center. Transportation Sustainability Research Center, Berkeley, UC, 2015. http://tsrc.berkeley.edu/bikesharing. Accessed October 20, 2016.
43. Walkscore. https://www.walkscore.com/. Accessed July 31, 2016.
44. https://www.ers.usda.gov/data-products/rural-urban-continuum-codes/. Accessed October 20, 2016.
45. Cronbach, L. J. The Two Disciplines of Scientific Psychology. *American Psychologist*, Vol. 12, No. 11, 1957, p. 671.
46. Nunnally, J. C. *Psychometric Theory*, 3rd ed. McGraw Hill Series in Psychology, New York City, NY, 1967.
47. R Development Core Team. *R: A Language and Environment for Statistical Computing*. R Foundation for Statistical Computing, Vienna, Austria, 2008, http://www.R-project.org
48. John, F. *An R Companion to Applied Regression*, 2nd ed. SAGE Publications, Inc., Thousand Oaks, CA, 2014
49. Satorra, A., and P. M. Bentler. Corrections to Test Statistics and Standard Errors in Covariance Structure Analysis. In *Latent Variables Analysis: Applications to Development Research* (von Eye, A., and C. C. Clogg, eds.), SAGE Publications, Inc., Thousand Oaks, CA, 1994, pp. 399–419.
50. Satorra, A., and P. M. Bentler. Scaling Corrections for Chi-Square Statistics in Covariance Structure Analysis. *Proc., American Statistical Association*, American Statistical Association, Alexandria, VA, No. 1, 1988, pp. 308–313.
51. Byrne, B. M. Structural Equation Modeling with EQS and EQS/Windows: Basic Concepts, Applications, and Programming. SAGE Publications, Inc., Thousand Oaks, CA, 1994.
52. Byrne, B. M. *Structural Equation Modeling with LISREL, PRELIS, and SIMPLIS: Basic Concepts, Applications, and Programming*. Psychology Press, London, United Kingdom, 2013.
53. Lavieri, P. S., V. M. Garikapati, C. R. Bhat, and R. M. Pendyala. An Investigation of Heterogeneity in Vehicle Ownership and Usage for the Millennial Generation. *Transportation Research Record Journal of the Transportation Research Board*, 2017. 2664: 91–99.
54. Hoyle, R. H. The Structural Equation Modeling Approach: Basic Concepts and Fundamental Issues. *Structural Equation Modeling: Concepts, Issues, and Applications*, 1995, pp. 1–15.
55. Kim, J., and K. H. Hahn. The Effects of Self-monitoring Tendency on Young Adult Consumers' Mobile Dependency. *Computers in Human Behavior*, Vol. 50, 2015, pp. 169–176.

The Standing Committee on Information and Communication Technologies (ICT) on Travel Choices (ADB20) peer-reviewed this paper (18-05902).

Sage Publications USA
sagepub.com

ISBN 978-1-5108-7838-9

CURRAN ASSOCIATES INC.
proceedings.com